The Beauty and Fascination of Science

Anatoly L. Buchachenko

The Beauty and Fascination of Science

 Springer

Anatoly L. Buchachenko
Department of Chemical Kinetics
Moscow University
Moscow, Russia

Translated by
Vitaly Berdinskiy
Orenburg University
Orenburg, Russia

Translated by
Katsuya Inoue
Hiroshima University
Higashi Hiroshima, Japan

ISBN 978-981-15-2594-0   ISBN 978-981-15-2592-6   (eBook)
https://doi.org/10.1007/978-981-15-2592-6

Based on a translation from the Russian language edition:
От КВАНТОВЫХ СТРУН до ТАЙН МЫШЛЕНИЯ by Anatoly L. Buchachenko
Copyright © LLC LENAND 2017
All Rights Reserved

This Springer imprint is published by the registered company Springer Nature Singapore Pte Ltd.
The registered company address is: 152 Beach Road, #21-01/04 Gateway East, Singapore 189721, Singapore

To my granddaughters
Anna Lapsheva and Nika Buchachenko,
and the generation of teenagers
meeting the charm of Life

Foreword

Each book has an author, and his name is on the cover page of the book. The author is also inside the book—inspired, faithful, and conquered by its beauty and fascination. He has the talent to transfer this beauty to the reader to capture his attention and to share with him his own delight. The author is writing delightfully about mathematics, physics, chemistry, and biology; he claims that exactly these sciences are indeed social ones, i.e., they serve mankind and each person, rather than to pretend to do that. I certainly recommend reading this book.

It is a marvelous story about science, mathematics, reigned science, and chemistry, which is not the whole life, but the life is totally chemistry about the charming biology, the mind, consciousness, and quantum strings. It is about everything remarkable in science.

Joint Institute for High Temperatures   Vladimir Fortov
Moscow, Russia
March 2019
L.D. Landau Institute for Theoretical Physics   Isaac Khalatnikov
Moscow, Russia
February 2019

Preface

To Reader
Books as the letters

> *…were written differently:*
> *Lacrimal, disease,*
> *Sometimes beautiful,*
> *Mostly useless.*

This book is about science, its inner beauty, charm, which becomes open through knowledge, through understanding. There is no other way. A French mathematician, the inventor of descriptive geometry Gaspard Monge: "only the charm that accompanies science can overcome people's disgust to the tension of the mind." And maybe not overcome. Because thinking is great work and elegant pleasure.

The author remembers the warning of Hermann Hess: "the world is teeming with writers who have a lot of great ideas, but there is no gift to find apt and bright words for their expression." A poet wrote about this in an easy way:

> *Limitless obscure visions*
> *Burnt their souls so strong,*
> *But could not get to people*
> *Through a twist of the tongue.*

The author shares the idea of a great Russian writer Mark Aldanov that the best of what was written on the tear-off sheets of the notebook was written without worries about publishers, readers, and descendants. This is how the author wrote these notes. But this does not give any reason to appreciate what he wrote.

These notes are about science and scientific creativity. About physics that is the science of the foundation of the Universe. About chemistry that is the central science in which we are all immersed and which has its own music and its own notes. And if chemistry is not all life, then all life is chemistry. These are notes about the delightful biology that is the science of the main thing, about life and the living world, about the mind, thinking, and genius. About mathematics that is the royal science. These sciences are the most humanitarian and the most humanistic. They serve man, rather than to pretend to serve.

The book is addressed to people with an open and curious mind—distrustful, but receptive. And age has nothing to do with it. But there is an author's focus on young minds, on those who make their choices, and who are looking for their ways in life and in science. Author's call before you go, before you step on anything, make sure it is not a rake. And remind Immanuel Kant: "…have the courage to live your own mind…" Russian poet Rasul Gamzatov puts it even more elegant: "Do not saddle other people's thoughts… Get your own." He was the son of the Caucasus and loved horses.

Moscow, Russia   Anatoly L. Buchachenko

Acknowledgements

My acknowledgements are to my dear and noble colleagues, the brilliant scientists Katsuya Inoue and Vitaly Berdinskiy. The great Springer Nature is worthy of special gratitude for its traditional benevolence to the authors.

Contents

About the Author and Translators

Author

Anatoly L. Buchachenko currently is the Professor of Chemistry and the head of Dynamics Department in the Institute of Chemical Physics of the Russian Academy of Sciences and the Professor of Chemistry and the head of the Department of Chemical Kinetics at the Faculty of Chemistry of Moscow State in University. He received his MS degree from Gorky Nizhny Novgorod University in 1958 and his PhD from the N.N. Semenov Institute of Chemical Physics in Moscow. He was a postgraduate student, Assistant Professor, Full Professor, and Director of ICP (1994–1996). Anatoly Buchachenko is a Full Member of the Russian Academy of Sciences. He was awarded the State Prize of the FSU for science (1977), the Lenin Prize, the highest award for science in the FSU (1986), the Voevodsky Award (1997), National Prize for Education Activity (2002), and National Prize Triumph for the highest achievements in science (2008). His scientific family includes more than 50 postdoctoral fellows, postgraduate students, and visiting associates. He is a member of the Advisory Board in *Chemical Physics Letters*, *Journal of Physical Chemistry*, *Russian Chemical Reviews*, *Russian Chemical Bulletin*, *Russian Journal of Physical Chemistry*, and *Mendeleev Communications*. Editor-in-Chief of *Russian Journal of Physical Chemistry. Focus on Physics*. Anatoly Buchachenko was born in Plesetsk now well-known Russian Space Center on September 7, 1935. He is married to Maya Buchachenko, technical editor of Science Publishers; their son Alexei is a Research Fellow in Moscow University, and their daughter is a book art designer. Hobby: modeling of the ancient wooden churches and wooden architecture.

Research Interest

His research interest includes chemistry and physics of stable radicals and high spin molecules, ESR spectroscopy, NMR spectroscopy of paramagnetics, spin chemistry, and organic (molecular) ferromagnets. Buchachenko is one of the founders and leaders in spin chemistry. He discovered magnetic isotope effect, a new remarkable phenomenon of fundamental importance for chemistry, geochemistry, physics, and biology. He observed microwave emission generated by chemical reaction (chemical maser, 1978) and formulated principles of chemical radiophysics. He is also the author of the radio-induced magnetic isotope effect and spin catalysis. He has contributed new ideas in design and synthesis of molecular ferromagnets.

Books by Anatoly Buchachenko

1. Chemical Generation and Reception of Radio- and Microwaves. A.L. Buchachenko, E.L. Frankevich. VCH Publishers, N.Y., 1994;

2. Spin Polarization and Magnetic Effects in Radical Reactions. K. Salikhov, Yu.Molin, R. Sagdeev, A. Buchachenko. Elsevier, Amsterdam, 1984;

3. Stable Radicals. A. Buchachenko. Consultants Bureau, New York, 1965;

4. Chemical Physics of Aging and Stabilization of Polymers. N.Emanuel, A.Buchachenko. VNU Science Press, Utrecht, Netherlands, 1987;

5. Chemically Induced Electron and Nuclear Polarization. A.L.Buchachenko (Moscow, 1974, in Russian; Bratislava, 1979, in Czech);

6. Complexes of Radicals and Molecular Oxygen with Organic Molecules. Science Publ., Moscow, 1984.

7. Magnetic Isotope Effect in Chemistry and Biochemistry. N.Y., Nova Science Publishers, 2009;

8. Chemistry as Music. Moscow, Nobelistica, 2004

Translators

Katsuya Inoue was born in Saga in 1964. He obtained his DSc (1993) degrees from the University of Tokyo. He became a JSPS researcher in 1992 (DC), 1993–1994 (PD) and got position of a Lecturer in the Kitasato University in 1994. In 1996, he moved to Institute for Molecular Science (IMS) as an Associate Professor. He has been Professor in Hiroshima University since 2004. Since 2014, he has also been working at Chirality Research Center (CResCent) in Hiroshima University as Director. In 2015, he awarded a Distinguished Professor, Hiroshima University. His main topic is development of novel functional magnets.

Vitaly Berdinskiy was born in Novosibirsk (Russia) in 1948. In 1972, he began to work as researcher at the Institute of Chemical Physics (Moscow, Russia). In 1983, he obtained his first scientific degree (PhD) for his work on chemical radiophysics. He worked in the Institute of Problems of Chemical Physics till 2006. Then he was invited to Orenburg University (Russia) as a Professor of Physics and head of the Department of Biophysics and Solid State Physics. In 2019, he was awarded the honorary title of Professor Emeritus of University Education by Russian Ministry of Science. His main topics are magnetic and spin effects in physics, chemistry, and biology.

Science

What Is Science?

That is boring. There are dozens of answers to it—bright and dull, serious and humorous, deep and primitive. Many of them demonstrate wit, sophistication, and elegance of thinking (like this: science is a way to satisfy curiosity at public expense). But accurate and indisputable is only one, simple and devoid of pathos—science is the extraction of Knowledge. Behind it all there is a goal, and methods of production, and ways of knowing, and profession, and inspiration, and fate. The history of science is a history of search and discovery, mistakes and successes, a way of overcoming dogmas… It is a movement along the roads of great, brilliant ideas and humiliating delusions, inspiration and despair, ups and downs, bright insights and dull, dull dead ends. This road is the road of crazy joys and lucid suicide, the road to delight and fatal errors. Great and dramatic roads of knowledge, endless and full of charm…

The great Newton concluded this idea in a monumental formula: "Science is the movement of human thought after the thought of the Creator." This formula is incomplete. First, Newton and modern science put different meanings into the concept of the Creator: he meant God, and science—the Big Bang. Second, Newton's formula is exactly depicted on his tombstone: "The purpose of science is about to follow God's thoughts towards a scary uncertainty." The end of the sentence warns that the movement to a frightening uncertainty. And this is a warning about the dangers of science. But more on that…

All collected on these roads are in books. And a little is in this… Books are compressed fuel of the future. But they can be a real fuel of the cave hearth.

© Springer Nature Singapore Pte Ltd. 2020
A. L. Buchachenko, *The Beauty and Fascination of Science*,
https://doi.org/10.1007/978-981-15-2592-6_1

In today's way of life, it is not so incredible. That is why it makes sense to look for inspiration in the beauty and charm of science—the highest and most mysterious part of our civilization. Turning the secret into the obvious, unpredictable into the inevitable–this is the intrigue of a delightful game called science.

Two Faces of Civilization

There is no higher and more complex challenge to intelligence and imagination than the one that sends science...

Two completely different worlds live in the brain, in this beautiful, delightful-divine organ, packed in a skull. One world is mind—the ability to reflect, to create a thought—the tool of knowledge; the other world—the spiritual, the ability to give birth to feelings, emotions. The first is objective, gives birth and holds knowledge; the second is absolutely subjective, individual; he is the Keeper of taste, likes, and attachments.

Science is just and fair—and this is also its charm... Newton's laws, quantum mechanics, dislocations in the crystal, isotope effects, the movement of planets—all these things are autonomous, independent of personal tastes; they are from the mind. You can love or hate them—they will not disappear, you will not get rid of them. Once young and future Nobel Prize laureate Nikolay Semenov (NN, all so-called him)—already the great scientist, but not the Nobel Prize winner yet (it would happen later)—standing at a board with chalk in hands, stated a certain new idea during a hot discussion. When someone from the audience noticed that the idea did not pass because it was contrary to the law of energy conservation, NN, thinking for a moment, agreed and threw with annoyance: "Oh, if not this law, how much good could be done!" He is right...

The world of art as a spiritual world is a subjective one. Someone likes Salvador Dali, and someone thinks him disgusting. Someone is crazy about the music of Johann Bach and someone indifferent to him and loves rock. For someone Russian writer Dostoevsky is a great thinker, but someone seems his philosophy as a primitive and cheap one. Someone loves ballet, and someone cannot see it. For someone, art is everything, while others consider it only a pale reflection of life. For someone, the cathedral Basil the Blessed in Moscow is a masterpiece, and for someone it is splint. The first people surely consider the second people as persons with poorly developed taste, people pathetic, deprived of aesthetic sense, morally deaf, and blind. It is unlikely that this confidence is reliable… Russian writer Vladimir Nabokov cannot be recognized as a man of bad or undeveloped taste, but here is his opinion of Dostoevsky: mediocrity, his books are full of melodramatic nonsense, cheap mysticism, and tormented suffering. And he is not alone in this assessment… However, a French archaeologist, Melchior de Vogüé wrote about Dostoevsky: "… a collector of the Russian heart, who was able to plunge into the sorrow of life." Famous Russian jurist, judge, politician, and writer Anatoly Koni said about Dostoevsky almost the same: "the poet of the mournful parties of human life." There were many critics of the beautiful poems of Russian poet Nikolay Nekrasov. However, Borovikovsky answered the most furious one of them:

> *You counted the sunspots,*
> *But overlooked his rays…*

The author knows many people who believe that everything written by Lev Tolstoy is not worth one page of what Svetlana Alexievich wrote in the book "The Unwomanly Face of War: An Oral History of Women in World War II." The described suffering of Andrei Bolkonsky is nothing compared to the suffering of those about whom Alexievich writes. The aristocratic grandmother of Russian poet Alexander Blok hated Tolstoy's moral sermons for their hypocrisy and banality. The theater is the great cultural phenomenon, but something false in life (e.g., gestures) is often called theatrical.

Of course, Don Quixote is a great hero. But one implausible circumstance confuses in his image: how can you live to such an advanced age and remain so chaste, he was naive, if not say—stupid? It seems that Cervantes portrayed the parody, which made the idol, attributing his fake nobility. But Nicaraguan poet Ruben Dario sees otherwise: "a warrior whom no one has yet been able to defeat, for his shield is a fantasy, and the spear is the heart itself…" For a Spanish-speaking poet, this is the truth.

Famous Russia architect Vasily Bazhenov is recognized as a great master of architecture (the Pashkov house in Moscow). However, writer Karem Rush considers him a man whose claims far exceed his modest inclinations; he is obsessed with unhealthy gigantomania and rebuilding on borrowed manners. Where is the truth? In art, in the world of feelings, there is no one truth; everyone has his own truth. In the evaluation of art, as in the estimates of people, there is no certainty. Only in science truth is the only one. This is the monumental difference between science and life: in science, everything must be right, and in life, people need everything to be good and not necessarily right.

In the spiritual world—in the world of art–estimates depend on positions, points of view, tastes, likes, and attachments. So they are diverse, often unexpected, stunning, discouraging, and exciting. There are many enchanting legends; one of them is the smile of Mona Lisa; there are millions of such smiles on the streets of Moscow… In the scientific world, assessments are objective and therefore boring and monotonous. This fact disappointed romantic nature of the Nobel Prize laureate Nikolay Semenov (see above). Well-known surgeon Nikolai Amosov sadly said: "for scientists science is primitive, and for others boring."

In the field of spirit, art, and culture, all assessments are ambiguous, tastes are changeable; what yesterday seemed aesthetic, today may lose its charm. Eternal values are rather exceptions; more often they are the product of firm beliefs, which are known to be the highest form of stupidity. Immanuel Kant claimed that creations of art are eternal ones only. It is necessary to forgive him this absurdity because he was a philosopher, not a scientist… Eternal values are really eternal science only and do not depend on beliefs.

There is a prayer passion in affection to a theater, painting, literature, and music; there is a fire of spiritual worship and religious delight. There are no such religious leanings in science, although science creations (from an automotive carb to a laptop) are of no less delight than the sonatas of Bach, waltzes of Chopin, or theatrical performances of famous Russian theater director Roman Viktyuk. A charm of those and others is available to different consciousnesses. And to different creatures…

Cognition is the most trembling process. Mind and feeling are two magical tools, tools of knowledge. Both are private property. A mind is the most indefinite concept, the most unsteady and flexible, and the most plastic of all human concepts. It is human nature to understand, even if he does not want it—this is how he works; the process of understanding, reflection, and comprehension of the truth is gratifying to him. The great Russian physicist Yakov Zel'dovich believed this property is the property of spiritual need. Understanding is the most important thing, divine occupation. And for some persons, it is also a profession… As one witty person said, parodying the famous philosopher: I think—and I exist for this… All people are brothers, but not all—by their mind.

The measure of a civilization is the degree of elevation of the human world over the world of natural wildlife. This measure, of course, should be evaluated by the highest achievements of the human mind and spirit; civilization does not exclude disgusting human creations—wars, murders, lies, meanness, cave television. Although, it could be a matter of taste for somebody as well as any unscientific emotional assessment… Nazis doctors who carried out inhuman medical experiments on children confidently considered themselves to be civilized people… The mass murder of Jews in gas chambers and in Babi Yar, they are also considered acts of civilization. The two faces of civilization are also its duplicity and its hypocrisy.

Science is at the edge of civilization: it supports the sustainable existence of what has already been achieved (life without science is the way to the caves) and produces new knowledge. They are followed by new elements of civilization. A Russian writer Maxim Gorky has written in a letter to biologist Kliment Timiryazev:

"Natural science is the Archimedes lever that is only able to turn the face of the world to the sun of the mind." A scientist and explorer Karl Ernst von Baer evaluated science even higher: "Science connects all educated Nations and once, maybe, will unite them in one common state Union." The idea is naive, but noble one… The union of two worlds—science and art, science and culture, their interference, and their interaction creates what is called civilization. It has two faces: there are strength, confidence, and reliability on the first face; there are variability, volatility, a rainbow of opinions, and the game of feelings on the second face. Art and science are two battlefields: the first one is the battle of tastes, opinions, views; the second one is the battle of knowledge and ignorance…

Global Ideas

Values in science and in society are different… People measure progress with such accessible and familiar things and concepts as TV, car, internal combustion engine, mobile phone, computer, space flight, nuclear energy, GLONASS (Russian GPS) system… There are a lot of such things—amazing, stunning and already lost the charm of novelty. People are used to them and almost do not realize where they come from. Most believe that all benefits come from shopping centers. On this subject, Einstein spoke well: "it should be a shame for someone who uses the wonders of science enclosed in an ordinary radio, and at the same time appreciates them as little as the cow those miracles of botany that she chews." Why ashamed? Einstein's too cool.…

For science, all these "miracles of botany" are only secondary products, side effects and consequences of its fundamental ideas. They are only events, technical and technological breakthroughs committed on the shoulders of the great global ideas on which civilization is built. And they are quite a bit…

Science has discovered the structure of the world. It showed that the world is amazingly simple, but in this mysterious simplicity, there is an intriguing mystery. Science has discovered the drawings and laws by which the world is created. She found that the world is created by precise mathematical legislation—by formulas and equations with precise world constants. This legislation is a science constructed in strict and accurate ideas and theories.

Global, creative ideas have created great theories:

- Euclidean geometry, the theory of physical space (Einstein called it the triumph of thinking)
- Einstein's theory of relativity, which created a macroscopic concept of the structure of our world and the Universe
- Quantum mechanics, which revealed the structure, dynamics, and amazing properties of nano- and microcosm
- Darwinian theory of evolution, a living and evolving concept of life, carrying a column of modern science

All the other numerous intellectual breakthroughs of mankind—and the flight into space, and steps on the moon, and nuclear power—are just "engineering" consequences of brilliant, creative ideas… Because neither the landing on the Moon, nor the flight to Mars, nor the synthesis of transuranium elements had fresh creative ideas.

Physics

Physics Is the Foundation of the Universe

Indeed, everything made by mankind rests on a solid foundation—on the principles and discoveries of physics. But a start in life is given to them by chemistry with its materials and substances. The truth is banal, but always forgotten and almost unconscious. Biology is the science of the main thing, of life… There is a regal science—mathematics; it unites everything, but also creates its own mathematical world.

Physics is the leading science, all-powerful and brilliant. The leader of the natural science and foundation of technical civilization… All the great, familiar and not dazzling, came out of it… Electronics and electrical engineering, radio engineering, nuclear engineering, optical and photo equipment, information technology… The list is endless… All around, all phenomena are the subject of physics: the Coriolis forces that twist the cyclones and tornadoes, and affect the weather; electricity and magnetism; Earth and ocean; the semiconductors and the atmosphere; nuclear reactors and ionosphere; radio physics and solid state physics; atomic nucleus and photon echo; holography and radioactivity; Sun and neutrino… And many, many more…

Physics is always at the edge of civilization, everyday discoveries are made in it. Most of them go unnoticed. We know about them only in new products, new products are included in our lives through what is called technical civilization.

Physics is a fascinating science. It overturns all predictions. When the great Ernest Rutherford came to understand the nuclear structure of matter and the disintegration of the nuclei, he believed a myth the idea of using nuclear

© Springer Nature Singapore Pte Ltd. 2020
A. L. Buchachenko, *The Beauty and Fascination of Science*,
https://doi.org/10.1007/978-981-15-2592-6_2

energy, and Albert Einstein agreed with him (see the Fate of discoveries). When Einstein came to the concept of stimulated emission, to the prediction of forcibly emitted light, he was sure that the use of this light phenomenon for practice and military purposes is a myth. But now there are lasers and even for military purposes… Lasers are everywhere: for fusion, for metrology and medicine, for optics and information technology, for science and art. Tiny lasers in electronic pointers and powerful lasers hitting military targets; continuous and pulsed lasers, gas and semiconductor lasers, free electron lasers, and chemical lasers. Huge laser world…

There are lasers compressing a ball of solid hydrogen with a diameter of 3 mm to monstrous pressures of millions of atmospheres and heating it to millions of degrees. They stimulate the fusion reaction of nuclear fusion in a time of about 30 ps, i.e., trillionth of 30 fractions of a second. This is the laser fusion. But the physics of lasers is only a small part of physics … Physics is a head of everything; physics is used at every step. By the way, even a human step is physics and biophysics of muscles; physics of angular momentum, physics of inertia, physics of gravity, etc.

Two Great Theories

Great conflicts gave rise to great ideas and were accompanied by great discoveries. The first conflict goes back to Isaac Newton (1643–1727). Among his discoveries, the most brilliant is the discovery of the laws of motion (Newtonian mechanics, general and universal). The most popular is the law of universal gravitation: all bodies are attracted, gravitate to each other with a force proportional to their masses. And this force is activated instantly; the concept of transmission of gravitational interaction is absent.

Knowledge process of the world is irregular: there are dull periods where nothing major is important in science is not happening, and there are stormy periods when the opening of the coming avalanche. As for fishing, I noticed Brian Greene. Almost a 100 years after Newton the great Michael Faraday established a connection of electricity and magnetism. He discovered a great phenomenon—electromagnetic induction, the property of mutual transformation of electricity and magnetism. And it had nothing to do with Newton or gravity. So it seemed…

But there was an amazing man—James Clerk Maxwell (1831–1879); he was the first Professor of the famous laboratory, established in 1871 at the expense of William Cavendish, Duke of Devonshire; then Ernest Rutherford and future Russian Nobel prize winner Pyotr Kapitsa worked. Maxwell left humanity totally accurate, flawless, and mathematically complete theory of fields (though its clarity came later and only with the help of those who followed him). The theory gives a classical description of the fields (including light) with a striking accuracy of 10^{-340}%.

It follows from Maxwell's theory that electromagnetic waves (including light) propagate at a constant and universal speed—at the speed of light. And that's the limit—nothing can move faster. And the light can't stop; it can't catch up and move with it; then it would seem motionless. This is allowed in Newtonian mechanics but is absolutely forbidden in Maxwell's theory. This is the first great conflict—the contradiction of two exact theories, a sign that one of them has a Vice.

In the same year, when the great Maxwell died, another great genius was born—Albert Einstein. The divine relay race … Einstein saw something that no one had seen before him—and this is a sign of genius: to hit a target that no one sees. He realized that space and time—concepts that were universal, robust, and obvious in Newtonian mechanics—were not really like that. They are unusually changeable, dynamic; they make up the fabric. We are immersed in the fabric of space and time and this fabric is changeable—it is compressed and stretched. It's ductile and its plasticity is a function of velocity… Rapidly moving bodies are contracted in the direction of motion because space itself is contracted. Time for such bodies is also compressed: fast-moving clocks are slower than stationary ones. And for light time is stopped, it is forever young. Relict radiation was born at the time of the Big Bang (12–15 billion years ago) and came to us being the same newborns.

These phenomena are incompatible with Newtonian mechanics; they do not fit into the concepts of common sense (on common sense, see below), but they are absolutely exactly inscribed in Einstein's special theory of relativity. Moreover, the equations of this theory are absolutely exact and quantitatively confirmed by many experiments. For example, sub-nuclear particles muons moving in the accelerator ring live longer and the increase in their lifetime, measured experimentally, strictly coincides with the theoretical predicted by the special theory of relativity.

J. Hafele and R. Keating experimentally compared the time of cesium atomic clocks, flying about 40 h on an aircraft, and the same, but still hours on earth. The experimental difference in their readings (the moving clock was one millionth of a second behind the stationary ones), measured to a fraction of a percent, ideally coincided with the theoretical one.

In Newtonian mechanics, the energy of a body with mass m_0 and velocity v is $m_0 v^2 / 2$; in Einstein's theory, the mass depends on velocity:

$$m = \frac{m_0}{\sqrt{1-(v/c)^2}}.$$

If the speed of the body approaches the speed of light c, the mass of the body tends to infinity. It is known from school textbooks: the light cannot be caught up, because the mass of the catching-up body becomes infinitely large. Electromagnetic light wave (photon) has no rest mass; moving at the speed of light, it has no age. Time has stopped for it. Relict radiation has arrived from the depths of billions of years; this ancient and eternally young light continues to travel in the world. And if we are confident, reliably aware of the Big Bang from the distribution of chemical elements, the history of the young Universe (300,000 years after its birth) was discovered due to the relict radiation. It appeared when the electric plasma of electrons and protons created atoms and stopped "swallowing" photons. Having escaped from the captivity of the young Universe, relict photons, these eternal Wanderers, bring us its history.

Having approved a new view of space and time, the special theory of relativity gave two fundamental consequences. First, the famous eq. $E = mc^2$; it determines the equivalence of energy E and mass m, their mutual convertibility. Here the great achievement of civilization came—nuclear and thermonuclear energy; it was understood where the energy of the Sun and other cosmic stars come from. Secondly, the theory has established a quantitative limit on the speed of movement of everything: electromagnetic disturbances, light, and any bodies. Nothing can move faster than light; this is the speed at which the ultimate deformation of the fabric of space and time is achieved. Nowhere on…

And again there was a conflict with Newtonian mechanics. Recall that the first conflict was resolved by the fact that the Newtonian idea of static space and stable time was overturned and confidently replaced by a plastic fabric of space and time, the deformation of which reacts to the speed of movement of bodies in this tissue. The second conflict is the contradiction between the Newtonian theory of gravitation and the special theory of relativity. The first assumes that gravity is activated instantly; the second gravitational interaction is transmitted at a finite light speed c. In the first gravity is a property of bodies, in the second—a measure of interaction.

And again Einstein has found the harmony of two incompatible theories, creating a *General Theory of Relativity* (GRT) in 1915, exactly 10 years after the special theory. In this theory, the idea of the plasticity of space tissue is developed—an idea that has already become a support in the exact and reliably proven special theory of relativity. In GRT, a body placed in space deforms distorts its flat surface and this curvature, this curvature is gravity. Gravity is a measure of curvature: a large curvature corresponds to a large mass.

Gravity as a deformation of space is transmitted as a mutual disturbance of interacting bodies. Suddenly born mass bends space and this curvature spreads through the fabric of space as gravitational invisible waves (like visible waves on the surface of the water from the fallen stone). These waves are carried at the speed of light, not instantly, i.e., not by Newton. If the Sun had disappeared, the news would have reached us through the gravitational wave only after 8 min…

Again common sense resists such physics. We will try to overcome this resistance, having thrown bridges from common sense to GRT. By the way, it was on these bridges that Einstein came to GRT. The first bridge was built by Galileo. He knew that the motion of bodies does not depend on these bodies: all he dropped from the leaning tower of Pisa, moved the same way. Hence Einstein's conjecture was born: gravitation can be not so much a property of bodies as a property of space.

On another bridge we always walk without noticing it; Einstein noticed. With uniform rectilinear motion in a train or plane, the passenger (any of us) does not realize the severity. This feeling appears only in those moments when the movement becomes curved (the train goes on a curved path, the plane rises or falls). Hence the idea that gravity is a consequence of the curvature of space. From these simple observations was born a brilliant GRT. It is accurate and strictly proven. Four clear and impeccable evidences should be recalled.

1. Long ago, in 1859, astronomers discovered a clear and impeccable contradiction: the orbit of the planet Mercury does not obey the Newtonian theory. The plane of the orbit rotates slowly (processes), the rotation angle is 575 angular seconds (575/3600 = 0.16°) for 100 years. Newtonian mechanics predicts an angle of 532 s. This difference of 43 s intrigued astronomers for more than half a century, even before Einstein, who realized that his theory removes this intrigue. He realized that the precession of the orbit is contributed not only by the mass of Mercury (so it is considered in Newtonian mechanics), but also by the additional energy of the gravitational field—because the energy is equivalent to the mass ($E = mc^2$), as follows from the exact special theory of relativity. And then the calculated period of the Einstein precession has appeared to be exactly the same as the measured one, rather than calculated by Newton. It was a great triumph of GRT, not everybody conscious.

2. The theory of GRT predicts the curvature of space introduced by the mass of the Sun; in this curved space, the light running from distant stars experiences the gravitational attraction of the Sun and bends its path. How to prove and measure it? First, astronomers measure the exact position of distant stars with their telescopes when the Sun is far away from the roads along which light from these stars moves to earth observers. They then measure this position when light passes near the Sun. The latter is feasible only during eclipses when the Moon completely covers the Sun but does not mask the light of the stars. And this measurement was taken during the solar eclipse on May 29, 1919. It was possible to measure the deviation of light rays by the Sun by comparing the position of the stars in these two modes.

   According to the GRT theory, the deviation must be of 1.74 arc-seconds. Direct measurements at two points of the Earth—in a small town gathered in the North-East of Brazil and on the Príncipe Island in the Gulf of Guinea—showed an almost perfect match with the prediction of the theory: the curvature of light rays was 1.61±0.30 s. The announcement of this new triumph of Einstein was made in London on November 6, 1919, at a joint meeting of the Royal Society and the Royal Astronomical Society. On 7 November this was reported by the London newspaper "Times," and on November 10 "New York Times" placed the article "Light in the sky bends." The world has understood nothing in this discovery, but choked with delight. The unknown Einstein became world famous, the most popular man on the planet, a man of unprecedented glory. (Not to say that this glory was clear, but about this elsewhere; see the Fate of discoveries). Einstein himself took delight in peace, he was sure of the result. The main proof of GRT for him was already behind him: the precession of the orbit of Mercury was an event that almost did not affect anyone.

3. GRT predicts the deformation of both space and time in a gravitational field. The space deformation was mentioned above (the orbit of Mercury, the bending of the light path); now about time deformation. Already in 1976, it was experimentally measured that the atomic clock raised by a rocket to a height of 10,000 km, where the gravitational attraction of the Earth is negligible, ahead of absolutely the same clock left in the power of gravity on Earth, four billion fractions of a second. This result coincided with the prediction of GRT with an accuracy of 0.01%. The brilliant charm of a new triumph…

4. General relativity predicted a new, unexpected property of light: he is experiencing the attraction as all the other bodies. The light emitted by the star loses energy, moving away from the star—the source of gravity. The

loss of energy is detected as an increase in the wavelength of light—the red shift. It was not easy to measure it: the light emitted by the Sun experiences a red shift of 2.12×10^{-6}, i.e., about 10^{-4}%. The difficulty is that this small displacement must be extracted from another source of red shift, for example, the Doppler effect. Still, the gravitational red shift was measured by the light emitted by the Sun at right angles to the observation beam (for this light, the Doppler effect is zero). The value of the red shift was about 2×10^{-6} as GRT predicted.

But even more precisely, the effect of the gravity of light was measured for gamma rays (short-wave, invisible to the eye light). They were launched from a height of 22.6 m or, on the contrary, thrown up to the same height in the tower of the physical laboratory at Harvard. Travelers departing from Earth, the gamma rays overcome gravity and lose energy detecting the red shift. The incident rays have gained energy and were showing blue (short wavelength) shift. Both effects coincided with an accuracy of 10% with the prediction of GRT. This was done in 1960; later the exact experiment resulted in a 1% match with GRT.

The brilliance and charm of GRT are that it is a perfect theory, canonically accurate and rigorous. Newton's theory of gravitation gives an expression known to schoolchildren for the gravitational force F between two bodies with masses m_1 and m_2:

$$F = g \frac{m_1 m_2}{r^2},$$

Here r is the distance between bodies and g is the gravitational constant. Newton's law is the law of the inverse square of the distance. But, as Stephen Weinberg noted, this law is almost empirical, it is "adjusted to the answer," i.e., to the quantitative laws of motion of the planets. Instead of a deuce, you can take 2.01 or 1.99 or something else—the law will not change, it will only be less accurate. But in General relativity the square of the distance is absolutely strong; nothing can be changed, otherwise, everything will collapse. It is a solid construction; it is perfect and beautiful in its perfection. And it perfectly accurately describes the world in which we exist and everything that surrounds us.

It became clear that Newtonian mechanics is included in General relativity, as an integral element, as its particular limit, the most important for life on Earth. Einstein is always right, and Newton reigns only at low speeds (far from the speed of light) and at small masses (far from the cosmic masses).

Einstein is absolute, Newton is relative. And both are the pride of humanity... As far as they have revealed and shown us the enchanting beauty and delightful harmony of the Great world.

Quantum Charm

So-called quantum mechanics is the science of atoms, electrons, photons, quarks, and everything else that is smaller... Quantum magic, quantum mystery, and witchcraft, crazy science, quantum magic—these are the names of this science. It opened a new, previously unknown quantum world and it was a tremendous breakthrough in the microcosm, in the world of magical phenomena.

Quantum mechanics is the most accurate and mysterious science, where everything is wrong, but everything is true and more than that—absolutely accurate. And it's mathematically as rigorous and perfect as classical mechanics. The "Royal equation" of quantum mechanics—the Schrödinger equation—is absolutely accurate; no phenomenon or event has been found in the microcosm that diverges from its predictions.

Like Einstein's theories, quantum mechanics was born in conflict. The temperature distribution of the radiation energy of an absolutely black body in the long-wave region, reliably and accurately measured experimentally, did not fit into the equations of the impeccable theory of heat—the classical theory. On the night of October 19, 1900, Max Planck derived a theoretical equation that accurately reproduced the experiment. But the agreement was reached in a monstrous way: Planck accepted that the energy from the black body is not a continuous stream, and portions, discrete shares, pieces—quanta. The assumption is ridiculous, but the experiment was for him. Serious physicists did not believe it, remaining in perplexity.

Further oddities multiplied. Atomic spectroscopy (first hydrogen atoms, and then other chemical elements) showed that atoms emit (and absorb) only certain, strictly fixed energies. And only they are allowed, and all others are forbidden, they are absent. Why? After all, in classical mechanics, any energy is allowed, in any quantity. Strange, isn't it?

But the event is even more ridiculous. A beam of electrons falls on a plate with two narrow vertical slits, and on the screen behind the plate, and which glows when electrons fall on it, instead of the expected two bands, a lot of bands appeared. There was a diffraction pattern—exactly the same as if light passed through the cracks as a light wave. So, is the electron a particle or a wave? Then the plate began to run electrons one by one, one after another, at intervals. Diffraction is preserved. Then, the electron goes through both slits being the wave and the paticle as single object simultaneously (https://physic-sworld.com/a/the-double-slit-experiment/). Irrefutable reliability of paradox-ical result…

Then another strange property of the microcosm was discovered: it is impossible to measure accurately and simultaneously the position of the particle and its momentum (or energy). In this world, there is something that is not in our large, macroscopic world: the more accurately measured coordinate, the less accurately determined momentum (energy), and vice versa. This is called the uncertainty relation, and it is not in the usual Newtonian world.

And there is another magical oddity: if a pair of particles (muons or electrons) is born from one place and at the same time (for example, during the decay of an atom), even if they fly to different ends of the Universe, they retain knowledge about each other. If something happens to one of them, the second instantly reacts to it. This state is called entangled; quantum computers and quantum computer science are built on it.

The main weapon of quantum mechanics is the wave function. Its own physical meaning is elusive, it is unsteady as a mirage, it is still the subject of fruitless disputes. Quantum mechanics cannot be understood, it is possible to know. As noted by Anatole Abraham, a French physicist of Russian origin, it does not need to be adored, it is necessary to work with it. Physicists have understood this for a long time, and when someone tries to talk about the meaning of quantum mechanics and its postulates, they say to him: shut up and calculate. And the calculations are exact absolutely; all its predictions are confirmed with stunning accuracy and reliability.

Yes, quantum mechanics is not so, but it is the greatest achievement of the twentieth century, the support of a new, modern civilization, its new face and new breakthroughs—from atomic energy to the landing of man on the moon. Everything in the Universe obeys quantum mechanics. And it is developing according to precise mathematical laws. But, unlike Newtonian mechanics, these laws determine only the exact probability of events but do not dictate the event itself. In the world of common sense, a probability is seen as a sign of ignorance; in the quantum world, on the contrary, it is exact knowledge.

It's like an eagle or tails game: the probability ½ is known for sure, but it's unknown what will fall out… For the great Einstein, this probabilistic, quantum world was unacceptable… And he was far from alone in the search for meaning. Stephen Weinberg in his beautiful book "Dreams of the Final Theory" tells an episode. Meeting his colleague, he asked him about the fate of a young man, a theoretical physicist, who had high hopes but suddenly disappeared from the scientific horizon. A colleague shook his head sadly and said: "He was trying to understand quantum mechanics."

Both Einstein and Weinberg tried to construct other versions of quantum mechanics, in which either the probabilities would disappear (Einstein), or it would accept other, common sense interpretations (Weinberg). Neither one, nor the other succeeded. Weinberg gave up his attempts, making sure that quantum mechanics is perfect and that it cannot be changed without destroying it all. But after all, it is and is irreproachably perfect foundation of the Universe.

Common Sense and Quantum Mechanics

Our common sense and life experience say that bodies can have any energy, and all objects—either particles or waves; the third is not given. In the world of quantum mechanics is the third: the objects of the microcosm behave inconsistently and illogically—they are transformed from particles into waves and back. In addition, they are allowed only certain energy states and are forbidden to have any, any energy. And these magical properties are not fictions; they are all strictly confirmed by all the experience of life and experimental science. It is irrefutable reliability of the paradoxical result.

Einstein, this great classic, for the rest of his life did not accept the probabilistic spirit of this science. "Quantum mechanics makes a strong impression, but the inner voice tells me that it's not the same. From this theory, it is possible to extract a lot, but it hardly leads us to unravel the secrets of God…" and he understood that "the laws of the Universe cannot be divided into the theory of relativity and quantum mechanics" (his words).

And then he threw the famous phrase: "God does not play dice…."

Playing! After all, the whole essence of quantum mechanics is in transformation. The electron as a particle, once in the energy pit, is transformed into a wave and leaves the hole as an electromagnetic wave (tunneling of the captive electron). On the contrary, the light electromagnetic wave, falling on the metal, is transformed into a particle—photon, which transmits its momentum to the electron, knocking it out of the metal (photoelectric effect). The whole world of physical and chemical phenomena is evidence of the transformation of the wave into a particle and back.

Einstein was psychologically unable to accept the idea of transformation. "The more success quantum theory achieves, the more stupid it looks". This rejection has become a source of his agonizing anxiety. "I no longer ask if there are quanta. My brain is unable to comprehend the problem."

Yes, no one understands quantum mechanics; it does not answer the question of why it is so. It is beyond common sense; for naive knowledge, common sense is not the whole truth. Common sense is based on logic, and it works only with obvious, naive knowledge. The world is richer, smarter, and more beautiful than the limited illusions of common sense. A British physicist David Deutsch said this very precisely: "if we want to understand the world not superficially, but more deeply, it is not our prejudices, acquired opinions and even common sense that will help us, but modern physical theories, in which there is much more sense than in common sense."

The search for common sense in quantum mechanics is a fruitless exercise, the field of activity of philosophers. The philosophy of quantum mechanics has nothing to do with it, it is empty and barren. Quantum mechanics does not exist as an approximation to some truth (as, for example, the Newtonian theory of gravitation is an approximation to Einstein's theory of relativity), but as the truth itself. The world is as we draw it, quantum mechanics.

However, rebellious minds do not leave Einstein's attempts. The young and talented mathematician M. L. Kontsevich believed that adding non-linearity to Hilbert spaces would inevitably "break" quantum mechanics. But it already did and it didn't break...

Quantum mechanics is like love—...
Love cannot be understood,
Love cannot be measured...

Quantum mechanics cannot be understood either, but it can be measured... And absolutely accurately. It gave rise to a refined, elegant theory—quantum electrodynamics, quantum theory of fields and moving charges. The accuracy of this theory is the same as if the distance from New York to Moscow was measured with the accuracy of the thickness of a human hair. And it's not just an impressive image; it is 10^{-8}% accuracy behind it. It is with such accuracy that the electron magnetic moment calculated from quantum electrodynamics coincides with the experimentally measured one. And do not forget that this perfect theory, so perfectly accurately depicting the world, is born at the tip of the pen and at the tip of the thought—is not evidence of the power of science and its great charm?

Common sense is also a property of habit, a kind of prejudice, a fragment of the old truth. It applies only to a small part of our world and our thinking. We are used to the fact that the energy is continuous, and the body—particles. Let's try to change habits and accept that energy is discrete, and particles are waves. And at once the world will appear clear, harmonious, and exact... Einstein is right: it is wrong that there are two worlds—quantum and classical. One world. Just in the quantum world of quanta are great—they are reliably detected and measured. When the particles increase, the quanta decrease. For large bodies with whom we deal in the world available to us, the quanta of infinitely small and inaccessible dimension. The largest particle for which quanta and wave properties have been experimentally detected is the charming C_{60} ball molecule (see below). The world is one and it is always and everywhere quantum…

Standard Model

It is the alphabet that matter is made up of atoms, atoms are made up of electrons, protons, and neutrons, and protons and neutrons are made up of quarks. And if the substance is text, then the electrons and quarks are letters. Smaller quarks only neutrinos, but it's not letters. This is taught in school; this is called the standard model of the Universe. And it is beautiful and slim.

But it was constantly invaded by new particles that were found on the accelerators. Additionally to two quarks, u and d, four more ones were added—c, s, b, and t; muons and muon neutrinos appeared, families of peonies appeared, which give birth to muons, K-mesons and hyperons were discovered. A huge event was the discovery of antiparticles predicted by quantum mechanics. The number of particles grew; reference books became thicker over the years. And more and more often physicists remembered Isidor Rabi, a great physicist, who even at the first unexpected discovery of the muon (it's almost an electron with a mass of 200 electronic masses) asked jokingly puzzled: "and who ordered it?" The harmony and beauty of the system were destroyed; the standard model of point particles became unsightly.

Now, many generations of physicists, studying the effects of bodies and particles on each other, have realized and proved that in nature there are fundamental interactions on which the world stands. Four of them, it is, of course, a little more than those three elephants, which was the world of ancient people. (However, there was a fourth turtle; but about it later…)

1. The strong interaction is powerful and short-acting, it keeps quarks in an inseparable state inside protons and neutrons, and also holds protons and neutrons in the atomic nucleus.
2. Weak interaction; it keeps the nucleon groups in the nucleus and overcoming it means radioactive decay of nuclei with the release of γ-particles and fragments of nuclei.

3. Electromagnetic interaction is the most common in everyday life, existing between all the particles and bodies carrying the charge.
4. Gravitational interaction is the most familiar, constant; it holds everything on Earth and the entire Universe.

The scale of the energies of these interactions is separated by almost 40 decimal places. If we take the gravitational energy per unit, the electromagnetic energy on average will be 10^3 times more; the energy of the weak interaction is 10^5, and the strong—10^{39}. Each of the interactions has its own transmitting field, and each field has its own particles, carriers of interactions.

Strong interaction "glues" quarks with gluons. The weak interaction is carried out by weak gauge W- and Z-bosons; they were first predicted, and then, in the 80s of the last century, discovered experimentally. Their masses are 86 and 97 proton masses and approach the mass of the Higgs boson (see below). The electromagnetic field is transferred by photons of different energy (they can be invisible to the eye); they are elementary carriers of electromagnetic interaction.

So, all three interactions are predicted by the standard model and experimentally found their particles, carriers of these interactions. In today's world, which exists now—billions of years after the Big Bang—three interactions are energetically "broken" and autonomous. However, quantum mechanics (and this, as we saw earlier, is an absolutely accurate science) predicts that at the time of the Big Bang and until about 10^{-35} s after it, the three interactions were in unity. The quantum theory of strong interaction was called quantum chromodynamics, and it was preceded by the theory of quantum electrodynamics, which combined the special theory of relativity with quantum mechanics.

After 10^{-35} s, strong interaction was "broken away" from the other two—weak and electromagnetic; the latter was a single interaction, called electroweak. Its theory is well developed and is called the quantum theory of electroweak interactions. And only a fraction of a second after the Big Bang, when the Universe expanded and the energy fields cooled, the weak interaction "broke away" from the electromagnetic. So there was an evolution of fundamental interactions and their "autonomy" is the legacy of the Big Bang, which we got. And it, this inheritance, is precisely and reliably controlled by quantum mechanics.

But there is also a fourth interaction (this is the turtle!)—gravitational one; it corresponds to the gravitational field in which the whole world is immersed, which is accurate, reliably and evidently described by the general theory of relativity (see above).

Quantum Foam

The first three interactions strictly obey quantum mechanics. Since the world is sole and born in one act—the Big Bang, the fourth interaction should not be an exception. When physicists began to combine these theories and solve the equations of both theories together, they discovered a new catastrophe: when leaving the microscopic scale of space due to the uncertainty relation—the fundamental law of quantum mechanics—space fluctuations are accompanied by giant fluctuations in the energy of the gravitational field. But the fluctuations of the gravitational field are the fluctuations of the deformation of the fabric of space (remember: in GRT gravity is the degree of curvature of space). And the smaller the spatial scale, the more powerful the fluctuations of both space and energy. In the limit of infinitely small elements of space, the fluctuations of the gravitational field become infinite—the answer is ridiculous and insane. This set of fluctuations, this chaos of physics called quantum foam. And the foam that starts at the scale of space is 10^{-33} cm; this threshold was called the Planck length.

Thus, the equations of GRT and quantum mechanics—both are perfectly accurate—are incompatible, they are in a state of absolute antagonism. Instead of the quantum theory of gravity, we got quantum foam. And then it turns out that there are two worlds: the macrocosm of boundless spaces and infinite times and the microcosm of quantum mechanics. And they live by their own autonomous and different laws. And it is impossible to combine them in the framework of the standard model; it is "within," but more on that later…

Is There a Theory of Everything?

So, unification is impossible… So what? That was the question to the problem from the school joke. Most physicists have come to terms with this incompatibility. Moreover, it is found on the scale of Planck length, and then you cannot get closer, not to split up space. But fine minds are not satisfied: the desire for harmony and perfection is limitless. The idea of the unity of the world inspired Einstein in his search for a Unified Field Theory, in search of unification stubbornly unwilling to unite the theories of the macrocosm and the microcosm. Each of them is perfect and perfect, but the fact that the two perfections are incompatible gives rise to a vague suspicion that this is some kind of imperfection, there is something unsightly.

But it's not even about aesthetics. New objects—black holes—were discovered in the Universe. By size, they belong to the "objects" of quantum mechanics, and by the giant masses, they are clients of GRT. What's that supposed to mean?

Quantum Strings and String Theory

And then quantum strings were born... Where and why—about this see **Mathematics**. The theory of quantum strings (or super-strings) appeared first as pure mathematics: the Euler beta function perfectly described the results of experiments on scattering and nuclear transformations in strong collisions of particles in accelerators; it accurately described the strong interaction. But why? Where is physics?

Answers, as always, bring ideas... Two physicists, Nambu and Susskind, saw that if we abandon the point structure of elementary particles (and on this, we remember, the standard model is based!), and to accept that they consist of one-dimensional vibrating strings, their strong interaction exactly obeys the Euler function; the size of the strings is close to the Planck length of 10^{-33} cm.

String theory has come a long and humiliating way from utter disregard to recognition. Brilliant success and greatness of the standard model obscured the theory of strings. It took several decades to understand that string theory is the most general one, that it is universal and includes a standard model as a limiting case. Quantum string has a set of energies and types (modes) of oscillations. Each mode and each energy state of the oscillating string correspond to an elementary particle with a certain mass, energy, and spin (angular momentum). All the inexplicable and embarrassing variety of elementary particles, which existed in the standard model, received a single explanation and found a single origin. Here a few words as an addition: aesthetic perfection is a sign of a reliable theory.

But there was another advantage behind the strings: since the strings are of a finite size, not a point, as it was considered in the standard model, the principle of uncertainty is satisfied in the joint solution of the equations of GRT and quantum mechanics. And then the quantum foam disappears, the antagonism that separated them in the standard model disappears. A significant event happened—the unification of general relativity and quantum mechanics.

Science and civilization are driven by ideas. They are always unexpected, and often just crazy. Such, almost wild idea was introduced in 1919 by a Polish mathematician Theodor Kaluza. He suggested that there may be hidden new dimensions, inaccessible to touch and realize. Kaluza did not use the GRT equations for three-dimensional space, as Einstein did, but added a fourth dimension to them. In the solutions, he found already known GRT results, as well as a new, unexpected result—the Maxwell electromagnetic field. It turned out that the new idea connected and united such different interactions—gravitational and electromagnetic: gravitational waves are transferred by the usual three dimensions, and electromagnetic interaction is carried out by a new, fourth dimension.

Here verses by Russian poet Valery Bryusov are remembered… Two lines from his poem "the World of N dimensions":

> *Height, breadth, depth. Only three coordinates.*
> *Where is the way past them? The latch is closed.*

Now the regret of the poet superfluous: the bolt open!

Then the inevitable happened: string theory and the idea of additional, hidden dimensions found each other. And that was the beginning of the Great Unification. Today—and again at the tip of the pen and the tip of mathematical thought—the contours of the great and Universal Theory of Everything were born, theory of substances, and fields, and micro- and macrocosm. This theory unites both mechanics—classical (Newtonian) and quantum; moreover, in this new theory, they cannot exist without each other. This is the theory of the great union, to which Einstein stubbornly went. This is the theory of super-strings, in which electrons and quarks—these fundamental particles—are made up of loops of vibrating, oscillating fibers, super-strings. And all the properties of the world are determined by their properties and their behavior. Stephen Weinberg defines it as the final theory (S. Weinberg. Dreams of the final theory, Vintage Books, 1994).

But this, of course, is not the end of science; it is only the crown of physical reductionism (see **Reductionism**)—the desire and ability to understand our world to the bottom. The way back is a new spiral of knowledge, a new and deep understanding of the world, which now sounds like the high music of quantum strings. And it is the strings that are the letters of the alphabet of our world. You can now restore the text…

Of course, everything was not as simple and logical as it is presented above. The unification of GRT and quantum mechanics, the creation of the theory of folded spaces and hidden dimensions, the theory of quantum strings, the unification of the four fundamental interactions—it was the way of ups and downs, disappointments and delight, brilliant guesses and dead ideas.

There are already several string theories; they have also given new life to the ten-dimensional theory of supergravity. (Our world is eleventh dimension, so as not to forget the time dimension). They develop, shimmer, improve, and this process will necessarily lead to a unified theory, to an understanding of how elegantly our world and our universe are arranged. But even today, like 400 years ago, Galileo's words remain true: "here so deep secrets are hidden and so sublime thoughts that the joy of creative search and discovery continues to exist." And next to them—the charm...

High Music of Quantum Strings

Still, common sense stands behind each of us and suggests the suspicion: is all this true? Yes, there is no common sense in both quantum strings and folded dimensions, but this is the case when there is no sense in the most common sense. Both collapsed spaces and the existence of strings cannot be proved experimentally: the energies at which the strings can be knocked out of the particles are unattainable. Physicists have long known that nobody can build an accelerator for such energy. But... There are consequences.

First, the string theory involved the standard model as a particular, limiting variant. And thus it satisfies the principle of reductionism as the first criterion of truth (see **Reductionism**). It already has a high chance of the string theory to be true. Second, the theory predicts that quarks cannot be separated, isolated. They have a strange dependence of the binding energy on distance: the farther away they are, the more energy is needed to separate them. Indeed, isolated quarks never could be observed experimentally at any energies. Third, string theory combined all four interactions, i.e., it reproduced what was at the time of the Big Bang. The standard model couldn't do it. Fourth, string theory combined not only interactions but also found folded dimensions. Moreover, it predicted gravity as inevitable (recall that before the string theory, gravity was a fact that took place, but did not follow from anywhere). The calculations predicted that in the quantum theory of gravity one of the modes of string oscillations is a graviton—a particle with zero rest mass and spin $S = 2$. This is the quantum of the gravitational field, a particle that carries gravity just as a photon carries electromagnetic interaction. Graviton is a string with a size of about 10^{-33} cm and it is folded so much that its calculated stress value is 10^{39} tons; it is called Planck tension. The graviton is not discovered, but none of the physicists in its existence are in doubt. Philosophers doubt ... But what does it matter? Here is the assessment of string theory of one of its harsh critics—Sheldon Glashow, Nobel laureate, leading theoretical physicist: "there are questions that cannot be answered within the framework of traditional quantum field theory… They can be answered by some other theory, and the only other one I know is string theory".

Entangled States and Quantum Teleportation

In quantum mechanics (and, of course, in life), there is another magical property—to create mixed, entangled states in which the quantum bond between the particles is preserved, even if they are removed from each other by billions of light years. These states are called mixed or entangled. In them, the particles are not autonomous, they know everything about each other; one has only to "touch" one of them (for example, to measure or change its angular momentum, spin), as the other instantly reacts to this event. And no interaction, neither electric nor gravitational nor magnetic—it is irrelevant. And it is weird, it is crazy, it is beyond common sense.

Einstein, this brilliant mind to which quantum mechanics was "across," came up with an explanation for this absurdity. He believed that quantum mechanics as science is incomplete, imperfect and therefore does not know how to determine the exact position and speed of particles. That exact position and speed are not available, although they exist, and there are some parameters in quantum mechanics, which are not yet discovered—so-called "hidden parameters." And once you find them, everything will fall into place and quantum mechanics will become "normal," acquire common sense. This idea is known in science as the EPR paradox (by the name of the authors—Einstein, Podolsky, Rosen).

In 1964, the Irish physicist John Bell showed that this idea can be tested experimentally. And this was done by French physicists led by Alan Aspect in the 80s of the last century. They measured the polarization (i.e., the actual spin) of two coupled photons. Their source was excited calcium atoms, which emitted simultaneously, in one event, two identical photons; the spins of these photons were measured by remote detectors.

The results of the experiments were amazing. If the EPR is right, the readings of the remote detectors should be identical—because the photons are the same. But the experiment found that the detector readings did not match. And that means that one of the photons reacts instantly to what is being done to the other, that they both make up one whole, no matter how far away they are. In fact, in the experiments of the Aspect, the distance between photon receivers was 13 m; later, in 1997, the experiment was performed with a distance of 11 km, but the result did not change. Then, in this century, entangled photons were detected at a distance of 144 km; the result was reproduced. And this means that EPR is wrong, that there are no hidden parameters in quantum mechanics, it is accurate and universal. It is not local (i.e., it does not act only in the vicinity of any quantum event), it is universal and perfect.

Moreover, it turned out (experimentally proved) that the entanglement can be transmitted by a relay—to link the entanglement of one pair with the photon of another pair and transmit the entanglement along the chain. This is the legendary teleportation—instantaneous transfer of states and, consequently, the information contained in them. And this is a general property of quantum systems; it is already confirmed for three coupled photons.

It could be assumed that all this science is a fiction at the forefront of intellectual avant-garde thought if there wouldn't two circumstances. First, there is direct experimental confirmation of this science. Second, there are two impressive and practical consequences—quantum cryptography and quantum computing. The first is based on the fact that entangled states can be created and transferred, but cannot be reproduced, cannot be cloned. Inherent in the information they are unavailable for "strangers," it is unbreakable, it is secret. And it works: at the end of October 2007 in the parliamentary elections in Switzerland, the results of the elections were transferred to the center of counting votes on the quantum line, invulnerable to hacking and interference from the outside.

Now a few remarks about quantum computers should be done. In a typical computer, information is stored in bits that take values of 0 or 1. (That is why children's game of heads and tails gave civilization incomparably more than a highly intelligent and clever chess game.) A cell of information in a quantum computer is a quantum bit, or states; as an atomic nucleus or an electron having two possible spin values—up and down; as a superconducting ring in which current can flow in two directions; as a quantum dot or an artificial atom—a small fragment of a conductor or semiconductor. The principal difference between qubits and bits is that the states of qubits can be entangled with each other. This means that all system states can be changed at once due to confusion. In a classic computer, such an operation would require 2^N steps in an N-bit system. This provides unprecedented parallelism of calculations (quantum parallelism), and it serves as the basis for the power of quantum computers.

And it works… In 2001, on the basis of magnetic resonance imaging has built the first quantum seven-qubit computer on the nuclear spin of the molecules of crotonic acid. So far, the most difficult thing he can do is to decompose the number 15 into prime factors. But everything is still ahead…

We Swim in the Higgs Ocean

There are several oceans on Earth that are known. But there is an ocean, not mapped out, an ocean in which we swim and the Universe as a whole; it is named after the Scottish physicist Peter Higgs. First, where this ocean was born and when it appeared, it is an energy field, and then—what it does and how it is detected…

The Universe was born from a tiny piece of matter weighing 10^{-5} g (the weight of a speck of dust) in space, the fabric of which is folded into a ball with a diameter of 10^{-33} cm. The Big Bang (the act of birth of the Universe) generated monstrously huge energy; it is believed (and quite reliably) that 10^{-43} s after the explosion, the temperature was about 10^{32} K. In the limited space of the early Universe, the laws of quantum mechanics and the ratio of uncertainties in position and energy created giant fluctuations of the energy field, raging in a compressed space. Further, as the Universe expanded, both the energy and the amplitudes of its fluctuations decreased. At some point (it is believed that it happened 300,000 years after the Big Bang), a beautiful phenomenon happened to the field: it "froze," passed into a stable state and remained as a cold echo of the Big Bang. The phenomenon is similar to the cosmological phase transition—as in the condensation of water vapor into droplets; so formed a stable energy field—the Higgs Ocean. (For physicists, everything becomes clear if we say that the Higgs Ocean arises as a result of spontaneous symmetry breaking.)

As we already know, each field, each interaction corresponds to a particle which is a carrier of interaction. The Higgs field transmits its influence by a special high energy particle—the Higgs boson; its spin is an integer, and the energy is equivalent to the mass of 100–200 proton masses. It is believed that it has already been found on the Geneva supercollider, although there are doubters…

What does the Higgs field do? Its function is universal and is detected at every step. Einstein showed that any mass bends the fabric of space and the greater the mass, the greater the curvature. But in GRT there is no answer where the mass arises, how it is formed. And here the Higgs field comes into play: it interacts with quarks and electrons, creates resistance to their movement, their acceleration, forming the inertia of motion and body mass. Mass is a measure of interaction with the Higgs field; this interaction is the source of mass. The photon does not interact with the Higgs field, so its rest mass is zero.

Ahead—search and charm of the new mysteries and disappointments maybe… Now the Higgs boson is an important element of the standard model. Its strength and weakness are known (see above).

This Strange Dark Matter

It is not fiction, not fantasy... A hunch about the mysterious dark matter appeared long ago, in the 30s of the last century. The German astronomer of the galaxy noticed that the mass of the galaxy cluster, reliably measured through the centrifugal force (by the speed of rotation of the galaxy with a known distance from the center of the cluster to the rotating galaxy; the students of the 8 class are able to do it), was many times superior to the mass of the galaxy measured by their luminosity. And even then there was the concept of dark matter (i.e., matter that does not radiate, does not glow). It is assumed that it consists of particles with a mass of almost a thousand times the mass of the proton; they have no charge and interact very weakly with each other and with particles of ordinary matter. It is also assumed that they were born immediately after the Big Bang, when the temperature was high and massive particles could be born, which further cooled and preserved. It is possible to approach these particles through the composition of cosmic radiation. For theoretical reasons, it should be expected that weakly interacting dark matter particles in collisions can turn into ordinary particles—protons and antiprotons, electrons and positrons, neutrinos, etc. The probability of this process is small but still finite. Magnetic spectrometer "Pamela," installed on the satellite, showed that the ratio of protons and antiprotons corresponds to that the standard model predicts. However, the number of positrons detected far exceeds what is expected from this model. Unpredictable excess of positrons from an unknown source of physics tend to be considered as a signal from dark matter.

Three properties of dark matter (and dark energy) are known: it is uniformly "spilled" in the Universe; it causes the Universe to expand with acceleration; the density of dark energy is constant. Now it is time for the most diverse hypothesis about dark energy... And one of them is that dark matter (and energy) does not exist at all; to explain the expansion of the Universe we must look for others to blame... With dark matter—it is dark, but still ahead... The search continues.

Mathematics

Mathematics Is Regal Science

The best cure for idleness is math.
This amazing science refreshes the brain,
promotes appetite and fosters friendliness.
It unites people and animals into one mathematical
set consisting of algebraic meanings.
                    (From an essay of 5th grade schoolgirl)

There is a real miracle in the world; it is mathematics—divine regal science, the magic invention of people, created on the tip of the mind and on the tip of the pen—and goose, and steel one. This charming science contains the magical property to predict the unpredictable and to unite the unconnectable. It amazes magical, mystical ability to guess the properties and phenomena. All the fundamental physical theories of the Universe are contained in impeccably strict and mysteriously accurate mathematical formulas. And they came before the theories themselves. Differential and integral mathematics brilliantly describes everything, although it was not created to describe the world; it was born as a game of mind and logic of thinking.

Mathematics is a special science… All other sciences—from physics to psychology—study and discover what is in nature. Mathematics itself creates, invents his subject, his object and he is versatile… And this is magic…

Black holes are magic and mysterious objects of the Universe, were found first in mathematical equations, and then discovered experimentally from astronomical observations of interstellar gases captured by these exotic

© Springer Nature Singapore Pte Ltd. 2020
A. L. Buchachenko, *The Beauty and Fascination of Science*,
https://doi.org/10.1007/978-981-15-2592-6_3

monsters. The positron (antielectron) was discovered at the tip of the pen in the relativistic electron theory developed by Dirac. A ridiculous prediction followed from Dirac's equations: the existence of an electronic double—an electron with a positive charge. This mathematical absurdity was so alien to physics that Dirac's first wish was to ignore it; only mathematical aesthetics saved this result. What was the astonishment of physicists, when the positron was discovered experimentally...

The great Maxwell theory of electromagnetic waves was not soon adopted by science. The legendary Oliver Heaviside did not need to confirm the theory: he considered it "obvious truth" only for the beauty and elegance of its mathematical structure. And he was right: it cannot be wrong that has such perfect mathematics. About Maxwell's equations, the great physicist Ludwig Boltzmann enthusiastically said: "is not God wrote these letters?"

About the philosophy, which was a science a century ago—about natural philosophy—Galileo respectfully said that it was written in the language of mathematics. Gottfried Wilhelm Leibniz (1646–1716), the creator of differential-integral calculus, was sure that even music is defined by mathematics: "Music is the hidden arithmetic of the soul, which cannot calculate itself."

When Einstein formulated his great idea of gravitation and curvature of space, he was surprised to learn that there was already a mathematical form ready to infuse his idea into it, turning it into a general theory of relativity. This form is Riemannian geometry. As Stephen Weinberg noted, it gives the strange impression that mathematics was waiting for Einstein, although neither Riemann nor Gauss, who created differential geometry, had no idea why they did it and whether it would ever be necessary.

There is a theory of Lie groups in mathematics named after the Norwegian mathematician Sophus Lee and created in the nineteenth century. And suddenly, a 100 years later, it was discovered that the Li SU(3) group brilliantly, in exact agreement with the experiment, describes the structure of elementary particles. Steven Weinberg has admitted that when he and Huang needed to describe the behavior of matter at extremely high temperatures, they have found almost ready mathematics in the works of Hardy and Ramanujan, the brilliant Indian mathematician who created the brilliant methods of calculating the number π in the theory of numbers. Hardy himself was proud that his pure mathematics would never be used; it turned out he was wrong...

The great Jules Henri Poincaré created an abstract mathematical apparatus of the dynamic theory of chaos; half a century later, the famous Russian physicist Andronov found in this theory a ready mathematical apparatus for the radiophysics and the theory of self-oscillations in radio engineering systems.

Heinrich Hertz wrote with admiration that the mathematical formulas of Maxwell's equations "live their own life, have their own minds; it seems that they are smarter than us, smarter than the author himself...."

From the naive children's game of heads and tails was born discrete mathematics, computer science. Their large-scale applications in chemistry and biology were opened. Even Darwinian evolution and historical processes tend to be described by mathematical models.

And another remarkable aesthetic quality of mathematics: the beauty of physical theories, their harmony and elegance are found in simple and elegant mathematical structures, where everything is internally consistent in perfect logic. Physicists believe that the ability of mathematicians to do unknowingly what is then, much later, needed for physical theories is fantastic. Eugene Wiener called this property "incomprehensible efficiency of mathematics." Hence the special relationship to mathematicians as to mysterious people equal to the gods.

The strong interaction that "glues" quarks together, ensuring the stability of protons and neutrons, was an intriguing mystery to physicists. Gabriele Veneziano, analyzing the results of high-energy collisions of particles obtained at different accelerators, in 1968, unexpectedly discovered an amazing phenomenon: the beta function, discovered by the great mathematician Leonard Euler (1707–1783), brilliantly accurately described the experimental results obtained two centuries later. From this unexpected and magical circumstance came the theory of strong interaction—one of the four on which the Universe and Life are built. Moreover, the theory of quantum strings was born from it; the theory, which opened the way to the fulfillment of Einstein's dream—the Great Unification, to the unification of what seemed to be undivided—to the unification of the mysterious laws of the microcosm, absolutely accurate, though devoid of common sense, and the classical, age-old laws of the macrocosm. And this great union is on the road of mathematics...

In 1919, Theodor Kaluza (then unknown German mathematician) did a ridiculous thing, absolutely crazy from the point of view of common sense: he introduced the fourth dimension into the three-dimensional Universe. By introducing what no one saw, as no one suspected, into the equations of General relativity (and Einstein created it for a three-dimensional, visual world), Kaluza solved new equations and found that their solutions included Einstein's equations. But they suddenly appeared and new solutions that Kaluza was astonished to find the equations of electromagnetism discovered by Maxwell in the nineteenth century. Suddenly the realization came that the fourth dimension combined gravity and gravitational interaction with electromagnetism and electromagnetic interaction. The unknown and enchanting power of mathematics led physicists to this unexpected and great breakthrough…

Further—it is more… The theory of quantum strings, which came out of the theory of strong interaction, demanded the introduction of new additional measurements. It turned out that the Union of the four fundamental interactions is simple, logical, natural, and mathematically elegant if we introduce into the equations of string theory in addition to the three visible dimensions of six invisible, strongly folded dimensions in the scale of 10^{-33} cm. Moreover, the most accurate solutions of the equations of string theory showed that in approximate solutions anyone collapsed dimension was lost and in fact, our world contains ten spatial dimensions and one time. Our world is 11-dimensional…

String theory is based on mathematics. To many, it seems like mathematical madness. But it has done what seemed impossible—has combined two absolutely accurate theories: quantum mechanics and the theory of gravity. The world has become one, as it is in essence. Further, the mathematical madness of string theory created a new, unified theory of elementary particles, in which the standard model of point particles was replaced by the model of strings of finite sizes. And finally, from the equations of string theory implies the existence (again math!) a special particle with zero mass and spin 2. No one found it, but in the theory of gravity, it was already recognized as a graviton—an undisputed quantum, a quantum piece of gravity, as a particle that carries the gravitational interaction. Again, mathematics leads physics and physicists on the roads of discovery, on the roads of knowledge of the world, again it unites what was habitually considered not to be united.

In quantum mechanics, in this greatest intellectual breakthrough of the twentieth century, there is a fundamental mathematical concept—the wave function. And its meaning and the meaning of its measurement is ambiguous; (she's) it's got a lot of judgment, a lot of ideas and concepts—smart and beautiful. The meaning of the function remains elusive, but the physical results of its application are absolutely accurate and perfectly reliable, it has earned unlimited confidence. Among physicists there is even a joke: when someone begins to talk about the meaning of the wave function, he is told—shut up and calculate…

Only once in his great life, Einstein did not trust mathematics and then regretted it very much recognizing this distrust as the greatest mistake in his life. Obtaining the key equation of the general theory of relativity

$$G_{\mu\nu} = \frac{8\pi G}{c^4} T_{\mu\nu},$$

linking the deformation (curvature) of space (tensor *G*) with the gravitational field (tensor *T*), he was surprised to see that it predicts an unacceptable result: the Universe is not static, the fabric of space must stretch or shrink, the universe is alive, it "breathes." Which means it had a beginning and maybe will have an end… This turn of thought predicted (from math!) the equation of gravity was incompatible with conventional dogma: the Universe is immobile, immutable, stable, and strong.

To satisfy this dogma, Einstein changed the original equation:

$$G_{\mu\nu} + \lambda_{\mu\nu} g_{\mu\nu} = \frac{8\pi G}{c^4} T_{\mu\nu},$$

having introduced a new term (cosmological term), where λ got the name cosmological constant. With the new equation, the Universe has become stable. But the equation with the artificially introduced cosmological term, mathematically unsightly, was subjected by Alexander Friedman and Georges Lemaître to criticism, from which Einstein first dismissed. But after 12 years, Hubble discovered that the Universe is not static, it is *expanding* as predicted by Einstein's primary equation. If Einstein believed mathematics, he would not hide this fact and would become the author of another great discovery. This realization provoked Einstein to the last hour. Mathematics was smarter…

But that is not all… Much later the knowledge and realization have appeared that the cosmological term still exists and works. It introduces a negative gravitational interaction (a new, counterintuitive concept); behind it stands the Big Bang theory, the Higgs field, dark energy, dark matter, inflationary cosmology—all of which make up modern and booming global physics. That is a different story, but again Mathematics reigns…

Mathematics does not explain, it is a means, a way to the Temple of knowledge. It gave birth to cybernetics and computer science as a new powerful technology, a new rise of civilization. It reveals the piercing beauty of the World, not the worst part of which we make up.

There is a deep and beautiful mystery—all excellent (by R. Penrose's definition), accurate and reliable theories are mathematically perfect, aesthetically beautiful. The equations of Newtonian mechanics are very simple, accessible to schoolchildren, but they obediently follow the laws of motion of planets and all other bodies. The equations of the theory of relativity are also not too complex, they are known by students, the Universe obeys them. The equations of quantum mechanics control the microcosm—photons, electrons, atoms, molecules—and they are also easily assimilated by students.

But once mathematics was shocked… Somehow Gauss in dispute with Avogadro argued that the laws of mathematics are irrefutable. Then Avogadro burned 2 L of hydrogen in 1 L of oxygen and received 2 L of water vapor.

Thus he showed that 2 + 1 equals 2 and not 3. Chemistry went AWOL mathematics… Perhaps this anecdote was invented by a chemist.

Evolution

Darwin's Evolution Theory

This is the greatest and most brilliant theory in which humanity has found its history and meaning, its place in nature. It has become the universal law, the wise Constitution to which the living world is subject. It showed that this world is a structure without a designer, that it is self-governing. The theory of evolution is the supporting column of the majestic building of modern science. It has withstood the violent attacks of the Church and attacks of the philosophers (more on that later), and strict scientific scrutiny. Every attempt to refute or discredit it was accompanied by its consolidation and new evidence of its truth.

The theory of evolution has opened the mechanism of development of the living world—variability and natural selection. Being a purely phenomenological theory based on observations of wildlife and findings of paleontology, in the twentieth century, it acquired a molecular basis—genetics and molecular biology. Combining phenomenological theory with molecular genetics and developmental biology has created a modern synthetic evolutionary theory.

Life is a global and unique phenomenon. "Life is a self-sustaining chemical system subjected to Darwinian selection" (from a report by the American space agency NASA). Every word is the key here. This is a multifunctional system, able to do everything, including the knowledge of the world in which it is immersed, and self-knowledge. This is a system that is able to provide itself with everything necessary, the system is constantly changing. Evolution is a way, a style of its existence. Evolution has no plan, she's blind, she doesn't

© Springer Nature Singapore Pte Ltd. 2020
A. L. Buchachenko, *The Beauty and Fascination of Science*,
https://doi.org/10.1007/978-981-15-2592-6_4

cook structure or function under future (Yes, she did not know) and does not do anything in advance. She adapts to the outside, the dead world, following his whims.

The most beautiful thing in the system, called life, is the mechanisms of its variability. And, of course, these mechanisms are chemical and molecular. They work at all levels—from the transformations of individual molecules to the evolution of molecular assemblies (nucleic acids, enzymes, synapses, etc.). They provide for the life of cellular structures, cells, and their huge assemblies—up to the whole organism, no matter who it is—a microbe, a worm, a dolphin, or a person. The fate of man is first of all biology, and only then—the circumstances. All living beings are brothers in the genome; the whole living world sounds the same genetic notes. As music is born from seven notes, so chemistry is built on 16 chemical notes, and all mankind sounds music 25,000 genetic notes. Yes, all men are brothers… But not always in mind…

The mechanisms of chemical evolution are discovered, established and proved by the main biological sciences—genetics, biochemistry, molecular biology. Both mechanisms and these sciences are beautiful and delightful… "In biology nothing can be understood otherwise than from the point of view of evolution" (prominent Ukrainian-American geneticist and evolutionary biologist Theodosius Dobzhansky). It is an absolutely accurate and flawlessly rigorous Manifesto of life. By the way, Russian writer Alexander Radishchev was the first Russian evolutionist; long before Darwin he announced that all life is "brothers—moth, snake, eagle, lark, sheep, elephant, and man." Spontaneous evolutionist…

Darwinian Evolution Compressed Thousands of Times

One of the reproaches of Darwinism is the slowness of evolution, a large delay between genotype changes and their detection in the phenotype: this creates the illusion of non-obviousness. The proofs of the theory are scattered in the present living world (a striking example is the rapid evolution of the finches of the Galapagos Islands), and in hundreds of thousands of years and thousands of generations in the world drowned in centuries—in the fossils of animals. It is like a slow-motion picture, millions of years long. Is it possible to compress this movie thousand times so as to observe the evolution in the eyes of one generation of people? The answer is positive and convincing in its clarity.

In our world, there is a huge kingdom of bacteria; they have their own genome, there is a synthesis of DNA and RNA, proteins, and enzymes. They have everything that people have. But the life of a man lasts a 100 years, and the life of bacteria fits into the clock, so the generation of bacteria is replaced at breakneck speed, and evolutionary cinema is compressed almost billions of times. And this is not reasoning, this biomedical practice deadly war human bacteria.

An example. In 1959, in the practice of medicine the antibiotic methicillin was introduced that was a penicillin analogue. It successfully struck the microbe *Streptococcus Pyogenes* (pneumonia) that is fatal to a man. But 2 years later there was another species (strain) of these bacteria not susceptible to the antibiotic, and 30 years later this new species became the predominant carrier of pneumonia. The microbe evolved almost before our eyes, it changed only a few genes that previously encoded target proteins for methicillin. And changed genes began to synthesize other proteins, other targets, indifferent to methicillin. In fact, there is a new strain of microbes, which again will be a grueling war. It is Darwinian evolution, compressed into hundreds of thousands of times. The evolution of microorganisms provides the most reliable and accurate proof of the definition of life as a self-reproducing chemical system subject to Darwinian selection. More precisely and in short cannot be determined.

The evolution of microbes is dominated by three mechanisms. The first (discussed above) produces chemical modification of genes, which is accompanied by changes in target proteins. In the second mechanism, the modified genes produce such proteins and enzymes that destroy, neutralize antibiotics, and this is a strong evolutionary mechanism of self-defense of microbes. The third mechanism uses the regulation of genes that control the operation of the transmembrane pump; thanks to the activation of these genes, the microbe gets rid of the antibiotic, taking it out of the cell through the cell membrane. In this case, it is not necessary to change, modify their own genes; often there is a transfer of genes and even whole blocks (cassettes) of genes between bacteria (it is called horizontal transfer and it is carried out by plasmids—non-chromatic ring structures). In any case, it is the Darwinian evolution of the bacterial genome and it gives rise to new species of living organisms. And it's compressed in time.

All of the above is a view of the bacterium as an enemy. But the evolution of its genome is the path of the inversion of the enemy into an ally. In the genome of some bacteria, few genes were found (and even dozens of sets of genes), in which the nucleotide sequence is such that they can synthesize enzymes. And enzymes are able to synthesize antibiotics or related molecules. But these "doze" genes by some reason, do not work. One way to wake them up is to make a nucleotide modification in the vicinity of such genes; such a technique in genetics is available. Another way is to produce an artificial horizontal transfer of genes encoding the synthesis of antibiotics from one type of bacteria to another. Such genetic manipulation is a way to create bacteria—producers of drugs and other useful things for humans. In fact, modern biotechnology is based on the principles of genetic modification and artificial selection of genetically modified microorganisms, and it creates medicines that save people's health and lives no worse than prayers. Note, incidentally, that Darwin came to the idea of natural selection through artificial selection by studying the brilliant results of human activity in creating new breeds of animals and varieties of plants.

Evolution of Intelligence

At the dawn of humanity, all people were almost identical—their intelligence was limited, and the distribution—narrow. In the 30,000 years of modern humanity, the distribution of intelligence has greatly expanded: from the many individuals who believe that the Sun revolves around the Earth, to geniuses, intellectual avant-gardists such as Einstein and Wiener. This is curious and mysterious—because all people have a common genome, each person carries the same 25,000 genes. These are the genetic notes that make up the symphony of humanity. Yes, all men are brothers, but not all in the mind…

There is another point of view. The distribution of intelligence is still as narrow as in childhood. In other words, man's childhood and humanity's childhood are almost identical in intelligence. Wide distribution is created as a result of learning, cognition, education, and thinking. And it depends on the living conditions, the circumstances of existence and what is called luck. And which point of view is correct? The strategy of human development may depend on its choice. But now it has no strategy, no leaders, and no intellectual avant-garde. Humanity is divided into numerous communities, often managed by not the most intelligent individuals. It seems that this is also an element of biological evolution…

Anti-Darwinism

Darwinian creation immediately met with the fierce hostility of the Church: its far-sighted leaders saw the threat of religion. And this is true—because it convincingly destroyed creationism as the Foundation of the Church, it abolished the need for the hypothesis of the Creator (Laplace's answer to Napoleon's question about God). It undermined the dogmas of the Church and, ultimately, threatened her income. The clergy organized a campaign of persecution, resorting to traditional weapons of slander, attributing the Darwinian doctrine, the preaching of the origin of man from apes. In those ages, it worked on the dark people. The reviews of Darwin's book expressed contempt for the author; it was predicted that the book would be forgotten in a month. The main accusation: Darwin's views are against religion.

The criticism did not abate, ever, she is still alive today. There is a pattern: the more ignorant critics are, the more violent and aggressive they are. They do not want to hear the arguments of educated people who know the Geology and geological history of the Earth, paleontology, and paleobiology, molecular biology and genetics, developmental biology and modern population biology, ethnic genetics and its compliance with the current geography. All these mighty Sciences demonstrate many thousands of irrefutable proofs of the theory of evolution.

Philosophers are tireless. Even the dignity of a theory they know how to put her in reproach. For example, they attribute the reason for the success of the theory to "the magic of visualization, which has led humanity to a dead end" (Tchaikovsky, Bulletin of the Russian Academy of Sciences, 2010). Life is unthinkable without the Creator—this is their main argument. The Creator is—otherwise you have to believe that "a new building can spontaneously erect itself, it is only necessary to deliver to the construction site a few tons of brick, cement, wood, and glass" (Rupert Sheldrake). Another option: self-construction of the aircraft from scrap metal landfill. The third, the most popular: a monkey seated at a typewriter and randomly hitting the keys will never print the text of the novel "War and Peace."

All this, of course, is shouting and stupid demagoguery. It is appropriate to remind Cicero: "Ignorance is the night of the mind." Evolution created elements of life gradually, not all at once, by trial and error, leaving useful, surviving and losing useless, unnecessary. Evolution is an accumulating process, its principle is to create, preserve, and use. "Life in its development has never neglected what has already been built but built on top of what is. Therefore, the cell resembles an archaeological site" (A. Szent-Gyorgyi. Science 176:966, 1972. https://doi.org/10.1126/science.176.4038.966), where layers stacked on top of each other mark the stages of the million-year evolution of the living.

All evidences of this are in modern molecular biology. All living beings have the same DNA and RNA, almost the same polymerases and ribosomes, ATP synthases and all enzymes, almost 99% unity of genomes and many other structures, the same delightful music of genes. Moreover, there are "extra" details that evolution has created (for example, introns in the genome); in some organisms, they are, but do not function—they are not needed. This is the case when evolutionary construction has proved redundant. Evolution is blind, its actions are devoid of accuracy, and solutions are not optimal (see **Reductionism**). But the general development follows the vector of Darwinian theory.

Reductionism Is the Criterion of Truth: But Not the Way to the Temple

Science is an object of miserable philosophy that is diligently imposed from the outside. But it has its own internal philosophy—reductionism, the most effective, impeccable philosophy developed by science itself; it lays down strict criteria of accuracy and reliability. And this is a great principle on which a strong education is built, for it stands on the shoulders of science. Reductionism is a universal principle of science, a strict and impeccable criterion of truth in discoveries and ideas. He argues that the fundamental laws written on the lower floors of knowledge, on the lower levels of the hierarchy of sciences, must be accurately executed on the upper floors and steps. New knowledge includes the old as an integral part. The general theory of relativity did not cancel the classical theory of gravitation—it included it. It is only ignorant journalists could shout headlines: "Newton was wrong!" The special theory of relativity has involved the electromagnetic theory of Maxwell. Quantum mechanics did not destroy the classical, Newtonian mechanics, it absorbed it as its component part, as the limiting case of low speeds and large masses. The theory of electroweak forces became a generalization of the electromagnetic theory and the theory of weak nuclear forces responsible for radioactive decay of nuclei. The theory of quantum strings did not destroy the standard model; it gave it a new meaning and a new sound. Modern synthetic theory of biological evolution has not changed anything in the basic Darwinian theory, it has expanded and strengthened, adding brilliant molecular genetics and paleontological evidence.

There are no fundamental, independent, and autonomous laws governing the behavior of complex physical, chemical, or biological systems. There is no need and sense to describe the structure and functions of the brain by equations of quantum mechanics, but there is also no doubt that in this structure and in these functions nothing violates the laws of quantum mechanics. It is the foundation of chemistry and therefore biology. Remove it and everything will collapse because the lack of understanding does not negate its existence. Reductionism is not just a way of hierarchical thinking. More importantly, it is the principle of the structure of the Universe, its fundamental unification and monumental unity. The world is just like this: it is logical, simple, and clear because in its heart "knocks reductionism."

In science, the concept of fallacious and false is defined almost unmistakably through reductionism. If a new idea or result does not satisfy the principle of reductionism, if they do not include known knowledge, but contradict it, one should be wary and suspect the fallacy of both the idea and the result. No, not discarded, but subjected to a comprehensive test of scientific accuracy and strength. Reductionism is the only reliably reasoned criterion of fallacious or false science. This is a filter that sweeps advertising and unreliable. Torsion engines with an efficiency greater than 100% are lies. Medallions of immortality, a magnet for money, astrological predictions, quantum biology, and quantum medicine are all ridiculous nonsense. And the beautiful word "quantum" does not save those labels…

Steven Weinberg has presented such a situation. A person suffering from a certain disease is offered to get rid of the disease in two ways: by touching the sacred person of the king or by eating chicken broth. It is clear that a person will choose the second—intuitively, unconsciously, he will use the principle of reductionism and make the right choice. The confessors of pseudoscience, who speculate on ignorance and trust, usually rely on the words of Indian poet Rabindranath Tagore: "If we close the door to error, how and where will the truth enter?" But this is pure demagoguery, wrapped in a web of beautiful words.

Philosophers don't like reductionism, they want to keep the secret at all costs. There are two extremes here: either total disregard for reductionism (which, of course, is a sign of deep ignorance) or its elevation to the absolute, which is dangerous. Even Confucius warned: "to study something and not think about learned is absolutely useless. It is dangerous to think about something without studying the subject of reflection." Of course, in principle we can reduce chemistry to basic physics and quantum mechanics, and biology to chemistry. But there is no need because there is a danger of losing many valuable concepts, useful knowledge, and depth of understanding.

Chemical Reductionism

In chemistry, there are no own autonomous principles and laws in which there would be no quantum mechanics or physics, but there is something that is not in quantum mechanics. Surface tension, humidity, viscosity—these properties do not have a single molecule, they are inherent in chemical ensembles. In chemistry, there are coherent phenomena when chemical processes are organized, ordered in space and time, but they belong to the ensembles of reacting molecules. In physics, quantum beats (also a coherent process) refer to the states of particles (for example, an excited atom). But there are also coherent ensembles; they are the source of laser radiation.

Quantum mechanics has counted and determined for chemical properties of the simplest chemical objects—hydrogen and helium atoms, hydrogen and water molecules, and other elementary two- and three-atom molecules. It gave a clear understanding of such fundamentals of chemistry as a chemical bond, exchange interaction of electrons, energy surface of reaction. It successor quantum chemistry learned to count and predict the properties of polyatomic molecules, bringing the technology of such calculations to recipes (self-consistent Hartree–Fock field methods, density functional theory, etc.). It created recipes of calculations of huge ensembles. In them, the behavior of a small number of principal key atoms is described by quantum mechanics and all the huge set of secondary accompanying atoms by methods of molecular dynamics. This technology successfully and accurately simulates the natural (thermal) life of molecules. But a competent chemist would not count a DNA molecule with its millions of atoms. And it is not due to the complexity of this calculation, but due to its uselessness: it will not give anything to understand the structure and functions of DNA, and to assert and prove once again the accuracy of the Schrödinger equation and the absolute irrefutability of quantum mechanics is a small madness. Reductionism is impeccable.

From the point of view of canonical physics, chemistry is simple: indeed, it is only necessary to take into account the Coulomb repulsion between negatively charged electrons, between positively charged nuclei, and the Coulomb attraction of electrons to nuclei. And, as it seems to orthodox physics, there is nothing to do in chemistry; all chemistry is simply reduced to physics. It is a formula of primitive reductionism; it is false, illusory simplicity. First, chemical systems are many particle ones (many electrons, many nuclei), and the problem of three bodies in physics is not solved strictly. Secondly, in the Coulomb interaction between electrons, there is a part that depends on the electron spin (it follows from the antisymmetry of the full wave function (the

Pauli principle) because electrons are fermions—this is how the world works). This part of the Coulomb energy is called the exchange energy; it builds the angular moments (spins) of electrons and creates ferromagnetism—a property on which ferromagnetic substances are built, which form the basis of electric generators, radio, television, communications, washing machines, elevators, computers, and other countless benefits of civilization. Figuratively speaking, if the chemical world is built "on Coulomb," then civilization rests "on Pauli."

You may trust Richard Feynman, who noticed that "chemistry is the most complex physics." Only a great physicist could say such a thing, although this compliment is a strong exaggeration. In physics and in chemistry everything known becomes simple, clear, and transparent. The meaning of the game called science is: to turn the complex into a simple, unpredictable—into an inevitable.

Physics and chemistry are always going close. Chemists discover new substances, reactions, phenomena; physicists look for explanations of properties, give physical mechanisms of reactions, build physical models and theories of chemical phenomena, and often predict new ones… Chemistry cannot be reduced to physics, each of these sciences is beautiful, has its own subject, its methods of knowledge, its handwriting, its masters and geniuses.

Having passed all the way of reductionism, chemistry has come to new ways of creating complex, highly organized chemical structures, synthesis of polyatomic ensembles capable of performing new chemical functions. The combination of the principles of covalent and non-covalent chemistry opens the way to new perfect structures with new functions that surpass even what nature has created.

Biological Reductionism

Modern biology vigorously follows the path of reductionism, the path of the chemistry of the living. The start of this movement gave a terrific guess of James Watson and Francis Crick (both are Nobel Prize winners) on the structure of the DNA molecule—this magic molecule, the icons of the molecular biology of the twentieth century. The trajectory of the movement was granted amazing discoveries in all fields of the science of the living world. We know how the root grows, how ATP—the main energy carrier in the living body–is synthesized, how contract muscles, how genes control protein synthesis, how the immune and signal systems function, how memorization occurs, and how synapses work. All biology with its DNA and RNA, with enzymes and ribosomes, synapses and neurotransmitters—all this is chemistry, a kind, elegantly arranged, beautiful, and smart functioning. And if the chemistry is not all life, then all life is chemistry.

The continual advances in biology have inspired Francis Crick to proclaim that "the ultimate goal of the modern movement in biology is to explain all of biology in terms of chemistry and physics." This is an extremely clear expression of extreme reductionism. It is attractive, it calls for new discoveries. But it is limited, moreover, it is primitive. And as chemistry cannot be reduced to physics, so biology cannot be reduced to chemistry. Biology occupies the highest position in the hierarchy of sciences. Of course, molecular machines, in which the electrical potential is converted into mechanical motion of molecules, there is in chemistry, but it is episodic structures and phenomena. In biology, they have become highly organized, systemic, and functional. In chemistry, there are no such perfect devices as DNA polymerase, ribosome, and synapse. Although, everything that happens in these devices is pure chemistry.

In chemistry, there are coherent processes organized synchronously in time and space. But such a level of coherence such as in muscle contraction, which is provided by coherent synthesis and decay of adenosine triphosphate, or such as in the processes of memory and thinking, which are provided by the coherent release of neurotransmitters and coherent reactions of neurons and synapses, is unattainable for chemistry. Here, biology is exceptional and unsurpassed.

But if a biologist reveals a reaction that is not in chemistry, it is suspicious. Perhaps this will be true if it passes through the filter of reductionism, which is enclosed in the structure and energy of electronic shells, in electron-nuclear interactions, in the chemical concept of reactivity. Chemists, especially physic-chemists, know this very well.

Chemical Evolution of Molecules

The age of the Earth is 4.6 billion years; rocks with the age of 3.5–3.9 billion are available to geology and only some grains of zircon are found even older (4.2–4.4 billion years). Earlier of the breed is unknown. The simplest single-celled organisms (this was a living cell!) appeared about three billion years ago, and microorganisms with a multicellular structure were born about two billion years ago.

Darwinian theory of evolution is the greatest theory, carrying a column of modern biology, it is subject to all life on Earth. But there is nothing in biology that is autonomous and independent of chemistry. Impeccable biological "Manifesto of life," formulated by our compatriot Theodosius Dobzhansky: "In biology, nothing can be understood without the theory of evolution." But just as perfect chemical "Manifesto of life": "in biology, nothing can be understood without chemistry." The variety of faces and eyes, height, mind, voice, character, talent—everything appeases from genes, all from biochemistry. And there are the same reactions as in normal chemistry—the decay and formation of interatomic chemical bonds. At the heart of life is variability and natural selection, and variability is the chemistry of genes and proteins. Chemistry is the driving force and source of evolution, it is its leading. Recall: life is a self-sustaining chemical system subject to Darwinian selection.

The living world was born from the dead world through its revived, highly intelligent molecules. This process of the birth of large and smart molecules, molecular machines (enzymes, DNA, RNA, proteins) from simple, primitive molecules H_2, N_2, H_2O, O_2, NH_3, and CH_4 might be called the chemical evolution of "dead" molecules into "living" molecules. Chemical evolution in this definition has nothing to do with Darwinian evolution; the latter says nothing about the origin of life. As noted by our eminent mathematician, cyberneticist, and philosopher Y. A. Schreider, do not ascribe to Darwin the intentions of Alexander Oparin. (Russian academician, a prominent scientist, was the first who posed the problem and began experiments on the transformation of mixtures of simple molecules such as CH_4, NH_3, N_2, O_2, and H_2O into more complex molecules, which could become raw material for biomolecules of life.) There are no definite answers about the origin of life on this path yet, but a number of reasonable hypotheses exist, they are set out in the wonderful book of academician Erik Galimov (The phenomenon of life: Between equilibrium and non-linearity: Origin and principles of evolution. URSS Press, 2001).

It is not necessary to invest in the concept of "animated" and "much wiser" molecules the literal meaning; it is not more than poetic image. Molecules, however, complex they may be, can be neither living nor intelligent. The egoistic gene (a concept introduced by R. Dawkins) is simply a romantic image of the behavior of this gene, in which there is no personal egoism. Chemists and biologists are professional and know exactly that there are no spiritual molecules. Life appeared when complex molecules united and organized into ensembles when there were signaling and transfer of commands between ensembles, self-reproduction of biomolecules and their ensembles and colonies.

In molecular evolution, there is much more mystery than clarity. There is no doubt that the simplest building material for life came from chemistry:

water under light reacted with oxygen, turning into hydrogen peroxide, which could oxidize methane into alcohols and aldehydes. The latter could give sugars as well as react with ammonia, forming something prior to amino acids. There are ideas how nucleotides appeared from aldehyde and hydrogen cyanide, which further through polymerization on inorganic catalysts created RNA and DNA. All this is clear only in principle, but Dmitri Mendeleev said, each thing should work "in metal," in reality rather than "in principle." A competent chemist will "draw" the chemical mechanism of formation of any biomolecule—the only question is what mechanism nature used, accumulated, and chosen; namely nature rather than a "cool-headed" chemist.

There is no doubt about the self-organization of biomolecules in the ensemble: here nature used the principles of non-covalent chemistry. The key to self-organization is interatomic potentials. In intermolecular contacts through atom-atomic potentials, molecules are selected by the method of trial and error, the most stable, low-energy states, the most stable conformations and forms, and the most stable associations. Atomic potentials are the source of the "mind" of the molecules (again the image!) helping to find partners and employees to perform their biological functions. Signaling, recognition seems to be carried out through phosphorylation of proteins and coagulation of their molecules, but there is no clarity with the birth of biomolecules. And not every self-reproduction is the way to live. When a crystal grows, it reproduces itself: each new layer of atoms, ions, or molecules reproduces the previous one. But the crystal has no biological functions, no life.

Of course, the accidental appearance of self-reproducing DNA and RNA molecules in nature, delightful molecular machines of DNA and RNA polymerases, ATP synthase, and ribosomes are excluded. Nature went to them the way of gradual complication and improvement, and sometimes workarounds. And there is evidence: it is known that some peptides are synthesized without matrix RNA and even without ribosomes. And there are peptides (e.g., gramicidin S and tyrocidine), which are synthesized directly from amino acids in the presence of only ATP and magnesium ions.

The question of the origin of life not only has no unambiguous solutions, but most biologists do not cause any inspiration. After all, we are talking about what has already happened: it is as uninteresting as to deal with the prediction of the earthquake. There is life—it has happened, and we must take care that it does not disappear so that humanity does not destroy it with its monumental folly and greed.

So, reductionism in biology, as in chemistry, and in science, in general, is the criterion of truth. But this is not the way to the Temple, not the way to understand life. In 2006, outstanding genetics academician Sverdlov published an article in the "Bulletin of Russian Academy of Sciences," the title of which was the question: "Biological reductionism goes? What is next?" The pathos of the article is the wise recognition that the larger and deeper the knowledge about the structure and functioning of biosystems (at different levels of the organization), the further goes and blurs the answer to the main question: "What is Life?" Continuing to move successfully along the path of reductionism, we cannot ignore the return path—the path of integration, generalization, the path to a new, higher understanding of life. And the first milestone in this movement should be a living cell, the "atom" of life. Any deep and complete understanding needs to overcome both ways—both in the direction of reductionism and back. The way back is already vigorously mastered by biology: from protein to the genome (not from genome to protein); synthetic genome, synthetic cell—these are the steps in this difficult but intriguing way.

Wise Niels Bohr noticed once that the existence of life in biology should be considered as well as the existence of quanta in quantum mechanics. Neither one nor the other can be derived from all existing and available human knowledge. Multiple (and even passionate) attempts to propose theories alternative to quantum mechanics, in which it would be possible to get away from quanta, were unsuccessful. About life, the conclusion is still obvious. It can appear only on the road, opposite to reductionism, a path with an uncertain finish, but full of unexpected and fascinating discoveries.

Chemistry

Chemistry Is in the Center of Sciences

Chemistry is not the whole life...
But all Life is totally Chemistry...

Chemistry is a central science. There is neither false pathos nor poor professional "patriotism" in this statement. Chemistry is not loved, even feared. However, one cannot get away from chemistry; one cannot get rid of it. It is similar to gravity. Any person has chemicals within themselves.

Metals and glass, plastic bag and cement, asphalt and concrete, paint on canvas artists and clothing, paper and markers, mascara and nail polish, food and surgeon gloves, rocket fuel and gasoline, drugs and deodorants, green grass and flowering trees—all these are chemistry... Each of us is a huge chemical plant, a giant chemical reactor, where everything you need for life is synthesized. Despite some failures, they are surprisingly organized (see *Biology*). Memorization and memory, thinking and feeling are chemicals... The production of ATP (the main energy carrier in the body) is chemistry. Chemical interaction of ATP with proteins is the source of motion. This fact gives us freedom of movement, makes our hearts beat; all these are chemistry. Our appearance is the result of chemistry, hair color, and shape of the lips, growth, and gait, voice and character—all laid in the chemistry of genes in the genotype.

There is its own internal music in chemistry; music of electrons and nuclei is music of 16 chemical notes—atomic orbitals. Everything plays a role on the

© Springer Nature Singapore Pte Ltd. 2020
A. L. Buchachenko, *The Beauty and Fascination of Science*,
https://doi.org/10.1007/978-981-15-2592-6_5

chemical scene. Electrons, atoms and molecules use these notes (see *Chemistry as music*). Chemistry is the science of substances and their transformations. It works and creates molecules of substances. Chemists have already brought more than 20 million substances "to the people"—10 times more than ones were created by Nature. Chemists are mages and wizards. Their magic is based on knowledge. Chemists can make any molecule, any substance. They know how the transformation of molecules and the transformation of substances proceed, how electrons and atoms move. They are able to anticipate the properties of substances and know how to make substances having the desired properties. They are the owners of a huge field, called Mendeleev periodic table. They are able to combine atoms of any elements (for example, to bond the first element with the 92nd and get uranium hydride). They have access to the delightful beauty and charm of a special architecture, to the molecular architecture of the world.

Chemistry as Music

Chemistry as a fundamental science was formed in the early twentieth century, along with a new quantum mechanics. Chemistry is a quantum science. And this is the absolute truth: firstly, all objects of chemistry—atoms, molecules, ions, etc.—are quantum; secondly, the central event in chemistry is chemical reaction, i.e., rearrangement of atomic nuclei and transformation of electronic shells, electronic clothes of molecules—reagents into molecules—products—is a quantum event.

Three main elements of quantum mechanics formed a solid and reliable physical foundation of chemistry:

The concept of an electronic wave function describing the distribution of charge and angular momentum (spin) of an electron in space and time
The Pauli principle that organizes the electron energy and spin states, that places the electrons on atomic and molecular orbitals (wave functions)
The Schrödinger equation as a quantum successor of classical mechanics equations

The awareness of these three "whales" makes absolutely clear and transparent all the majestic, monumental building of chemistry; all the richness and diversity of chemistry is born of them, its harmonious logic, beauty, and perfection is laid in them. They transformed the periodic system of elements into a periodic law, which controls the filling of the electron shells of atoms and molecules and dictates the chemical behavior of atoms; it follows a coherent, rigorous and flawless theory of the chemical structure of matter. Of course, you can work as chemists and do wonderful things without knowing quantum mechanics and without using it. But then amazing thing disappears—a holistic perception and a sense of unity chemistry, awareness of it as a unified system. Without quantum mechanics, chemistry breaks down into pieces; each could be both beautiful and valuable. But the greatness and monumentality of the building of all chemistry can be seen only on the foundation of quantum mechanics. And it brings an amazing clarity of understanding of the simplicity of the world.

The Schrödinger equation is the key to all chemistry. Its "chemical" meaning is easy to see by solving this equation for the simplest particle—the hydrogen atom; it is from it that chemistry begins. The equation looks like:

$$H\Psi = E\Psi, \tag{1}$$

here H is the energy operator

$$H = \frac{p^2}{2m} + \frac{e^2}{r}, \tag{2}$$

which includes the kinetic energy of the electron $p^2/2m$ (p is the momentum, m is the electron mass) and potential energy e^2/r (e is the charge of the electron and proton, r is the electron-proton distance). The solution of this equation (it can be found in textbooks) gives discrete energy levels E and the electron wave functions (orbitals $\Psi(r)$), whose square $|\Psi(r)|^2$ gives the distribution of the electron charge and spin in space.

The $\Psi(r)$ functions are wave functions, *basic atomic orbitals*—they make up the electron shells of all atoms and molecules, all chemical particles, the entire chemical world, the entire Universe. The mathematical elegance of chemistry is created by the wave function $\Psi(r)$. There are only 16 of these functions: *one s-orbital, three p-orbitals, five d-orbitals, and seven f-orbitals*; the spatial orientation and shape of *s-, p-, d-,* and *f*-orbitals are shown in the figure.

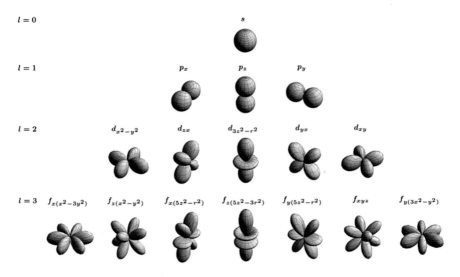

And as well as endless chess are born from combinations of simple chess moves, as magic and eternal music is born from seven simple musical notes, so powerful and inexhaustible chemistry is created from 16 simple atomic orbitals—"chemical notes." So the entire chemical architecture of the world is built. The science and art of chemical synthesis is the knowledge and foresight of how to combine the orbitals of atoms to produce their chemical assembly, creating given molecules and substances. These notes are played by all persons acting on the chemical scene—electrons, atoms, molecules, ions, clusters, etc.—creating the music of the chemical substances and processes. When atoms are combined into molecules, they combine their atomic orbitals. These now generalized atomic orbitals create common molecular orbitals—multi-electron wave functions; these combinations of atomic orbitals make up the chemical score.

When chemists create molecules, synthesize substances, control chemical reactions, they actually perform this chemical score—it is either already known and written earlier, if a known substance is made or a known reaction is carried out, or it is written, composed a new, if we are talking about new substances and new processes. In chemistry, it is like in music, or singing, or composing… Chemistry is music of electronic orbitals, but you need good musicians and singers, composers, orchestras, and conductors to make it sound. As in music, chemical notes have their own sharps and flats—hybridization and undivided electronic pairs; there are octaves—the main quantum numbers that define the "energy" sound of orbitals.

It is worth paying attention to one fact: in Eqs. (1) and (2) the potential energy of the chemical system electron + proton includes a single interaction—Coulomb one. All the basic atomic orbitals, these chemical notes, following from the solution of Eqs. (1) and (2), are built only on the Coulomb potential. Is it true? And don't we need to take into account any other significant interactions (just as in nuclear physics it is necessary to take into account several interactions of different nature)? The answer to this fundamental question was obtained experimentally. The idea of the experiment is simple and beautiful. A beam of atoms (for example, calcium or copper) is irradiated with polarized light; light excites, i.e., transfers an electron to a given atomic orbital (for example, p_z-orbital in Ca and d_{z2}-orbital in Cu, see figure), i.e., prepares atoms with electrons inhabiting these orbitals. Then a probing beam of inert atoms (for example, helium) is directed to this beam and scattering of these atoms on electrons inhabiting p_z or d_{z2} orbitals of scattering atoms is observed. By analyzing the scattering diagram, it is possible to reproduce the shape of the scattering potential.

The results of these experiments are amazing: it turned out that the shape of the scattering potential accurately reproduces the shape of the atomic orbital derived from the Schrödinger equation and shown in the figure. You can, of course, say that it is banal, that it was expected. But we must remember that the wave functions (atomic orbitals) are obtained at the tip of the pen and at the point of thought—from the Schrödinger equation with the Coulomb potential. So, two fundamental conclusions follow.

First, the Schrödinger equation correctly describes the chemical world (and hence the whole world); there is no fact or phenomenon of the physical or chemical world that contradicts this equation.

Secondly, experimental proof of atomic orbitals is not just a triumph of intelligence; it is also reliable evidence—only Coulomb interaction works in chemistry. Thus the magic of chemistry and its charm lie in the fact that from this Coulomb uniqueness a wealth of chemical bonds is born, as well as a variety of chemical particles, a wealth of mechanisms of reactions and chemical states.

Chemistry Is a Social Science

This is a humanistic science… It does not just create beauty; it also knows how to benefit from it, to satisfy the interests, needs, and whims of society and each person. It's addressed to everyone. Chemistry and chemical culture have become an integral part of civilization, an element of sustainable development of society.

Health and medicine. Health and intelligence are two blessed things; this was understood by the ancient sages. Both are chemistry. Physical health is based on the coordinated, harmonious work of many thousands of enzymes catalyzing chemical processes working in the giant chemical reactor, which each of us carries in itself. Intelligence is the result of the biochemical activity of the nervous system and brain; the processes of transmission and storage of information in molecular structures and chemical reactions; the processes of thinking, generation of thoughts, and ideas—all these are chemistry. All neurophysiology is neurochemistry. Emotions, passions, moods, emotional movements, and impulses—behind these is the chemistry of feelings, the chemistry of passions, behind these are molecules and their chemical life.

Any disorders in the work of enzymes and deviations from their coordinated work are a disease. The great destiny of medicine is to correct defects in the chemical life of the body, ridding it of diseases. But medicine without chemistry is helpless. Chemists armed medicine with a huge range of medicines that are isolated from plants, animals, and microorganisms and are synthesized in chemical laboratories and industrial reactors. Many of the drugs made a popular joke: "Tell me, doctor, do you have a cure for my disease?—Yes, we have so many drugs that you don't have so many diseases."

Like any aphorism, this joke is false. There is no cure for cancer, there is no reliable remedy for murrain and mad cow disease; the flu is so many-sided, its virus is so diverse and changeable that chemists only pursue it, never catching and killing once and for all. Chemists are aware of their social responsibility and are actively looking for the legendary "panacea" for these diseases.

These vast territories have already been conquered by chemists in alliance with doctors. A huge number of regulators of enzymes, inhibitors, receptors, mediators, and substances have been found that can treat hypertension, atherosclerosis, asthma, stomach ulcers, etc. It is worth recalling antibiotics and a variety of antiviral drugs that protect the body from viral infections. Drugs are found that regulate the cardiovascular system, saving from hypertension, atherosclerosis, arrhythmia, heart failure, regulating blood pressure, strength and rhythm of the heartbeat. The artificial "driver" of the heart muscle runs on a tiny lithium battery that is the invention of chemistry and chemists. Implantable batteries with a service life of more than 10 years have revolutionized medical biophysics; they have extended the life and happiness of hundreds of thousands of people.

The decisive contribution to establishing the nature of the molecules of the antibody belongs to chemists; they showed that this protein identified their chemical structure and the structure of their encoding genes. Chemists identified hormones that regulate the cycles of reproduction, established their structure and functions, opening the way to cure diseases that violate repro-duction. Great success has been achieved; many drugs and drugs have been created that perform stimulating and regulatory functions in a complex and delicate reproduction system. It is only appropriate to note that the illiterate, unprofessional possession or abuse of these funds is dangerous, but there is no fault of chemists: you cannot lose your head just because there are headaches.

We have seen huge breakthroughs in genetics; chemistry and chemists again play a major role; in fact, all modern genetics is part of chemical science. It promises huge breakthroughs in medical technology, in multiplying the benefits of life for a person, but it also carries with it major dangers. The "ben-efit/harm" ratio is again a question of the morality of those who use the achievements of genetics.

Chemistry is actively invading neurophysiology; hundreds of neurotrans-mitter molecules have already been discovered that affect the functioning of the central nervous system and, in particular, the normalization of the dis-turbed brain (Parkinsonism, Alzheimer's disease, and other pathologies). Here are some examples. Reduced levels of dopamine—a neurotransmitter respon-sible for coordination of movements was found in the brain and body of people suffering from Parkinson's disease. Chemists have found that the pro-duction of dopamine in neurons is suppressed by the protein A2A: it reacts with adenosine and the product of this reaction inhibits the enzyme that pro-duces dopamine. Experiments in mice found that caffeine doses inhibit the reaction of A2A with adenosine, i.e., the production of dopamine in the pres-ence of caffeine remains high. This is the way to drugs for Parkinsonism.

Another example: it was found that the gene called CHIP is responsible for human stress; it is involved in stacking, folding (folding) proteins. Under the influence of stress hormones CHIP folds proteins wrong: it spins the mole-cules and packs them wrong. Then proteins begin to "stick together," they are not destroyed and are not removed from the body. An excess of these cor-rupted proteins causes Parkinson's disease and Alzheimer's disease and stimu-lates blood clots and heart attacks.

The third example relates to the chemistry of memory. Chemists and genet-icists have known that NR2B protein improves memory and learning. Then they injected an extra gene that encodes and produces this protein into a fer-tilized mouse egg and implanted it into the mother mouse uterus. As a result, there were transgenic mice-children, sharply allocated mental abilities (easily trained, resourceful, etc.). Excess protein NR2B in the neurons of these "geeks" activated nicotinamide-methyl-D-aspartate receptors, which are responsible for associative thinking and development of the mind.

Even the layman is clear, what the new horizons in the treatment and in the improvement of man through the chemistry of genes, using gene replacement through transgenic engineering are appeared. This is a high and long scientific flight, but dangerous, concealing unforeseen troubles. The difficult task is not to cross that unsteady line when good turns into evil.

Chemists have created brilliant methods of chemical diagnosis of diseases (deviations from the norm of the chemical composition of blood, saliva, urine, sweat). Among them, nuclear magnetic resonance methods developed jointly by physicists and chemists are a real masterpiece; they allow to determine the chemical composition in samples taken in the amount of 10^{-9} g (nanograms) and in the amount of 10^{-6} cm^3 (microliters), to see the anomalies of the composition and to diagnose the disease.

Food and its production. Most toxic substances in food are produced by nature rather than in chemical laboratories. Microorganisms make much more dangerous chemicals than those used in agriculture. To prevent bacterial contamination, chemistry has created safe preservatives, means of sanitary perfect packaging and packaging of food, means of chemical control of food. Chemists found means of veterinary protection of animals, created tests of quality of milk, standards of purity of food. Leavening agents and other chemical additives in the production of bread give it excellent quality; fermentation and aging in the production of beer and wine are chemical processes and are accompanied by chemical control. Food supplements (including trace elements and vitamins) come from chemistry; they improve food and serve human health. Chemists have developed methods for cleaning and disinfecting water; this is huge merit of chemistry, which saved humanity from global epidemics, although this is rarely remembered.

Chemical "support" of food production, care of its quantity and quality begin with agriculture. Chemical fertilizers supply the soil with nitrogen, phosphorus, and trace elements necessary for photosynthesis and plant growth; hormones and growth regulators that stimulate cell division accelerate seed germination and plant growth, their vitality and endurance—all these are also the contribution of chemistry to providing mankind with quality food. Not all know (and those who knew have already forgotten) that boneless grape is also a gift of chemists.

Chemists have created plant protection products that are safe for humans but destroy weeds and insects (herbicides, pesticides, defoliants). Together with biologists, they have developed brilliant methods and technologies for the destruction of insects dangerous to agriculture and food. Pheromones are sex attractants that attract insects of one sex or another, which brought great success. Traps with these substances can save huge areas of fields and forests from pests. People know little about this noble work of chemists, but almost everyone knows about the troubles with DDT—dichlorodiphenyltrichloroethane. It is a powerful insecticide that destroys insect pests. DDT was put into practice in 1944, and the Swiss chemist Paul Müller received the Nobel Prize for discovery of this compound. However, a decade later, DDT was found to be persistent and slowly disintegrating. It began to accumulate in living organisms along the chain of plants-animals-man. Bird populations were particularly affected: DDT interfered with reproduction in such a way that the enzymes responsible for eggshell synthesis were suppressed. As a result, the bird offspring did not survive, the bird population felt. Immediately DDT was banned, but in people's memory, this story and prejudice to chemistry remained. At the same time, no one remembers that DDT saved humanity from malaria. Before DDT, malaria killed two to three million people each year. And in tens of times, more people suffered. 10 years after stopping using DDT, this horrifying tragedy is almost gone.

It should be noted that chemists are not enemies of humanity and themselves. They are guided by good intentions, but long-term consequences cannot be foreseen. And it is not only in chemistry, but it is also everywhere in politics, economics, and ecology. There is a joke: if I would be as smart as my wife later…

Fuel and energy. We must not forget that gas, oil, coal, and wood are of chemical origin; they are based on photosynthesis, the chemical process of storage and conversion of solar energy. Processing of oil and gas, production of gasoline and diesel fuel, asphalt and polymer materials—all these are chemistry. Combustion, i.e., the return of stored energy, is a purely chemical process, regardless of whether the wood is burning in a village furnace, gasoline in an automobile engine or solid fuel that propels a rocket into the sky. Chemists have developed combustion regimes with a high conversion rate of fuel energy into useful work, combustion regimes of high ecological purity (without the formation of dioxins and other toxic products), technologies of household waste incineration. They found the substances—additives, by which it is possible to change combustion modes. For example, an air-gasoline mixture in an engine ignited by a spark creates a combustion wave propagating in the engine cylinder at a speed exceeding the speed of sound. The temperature and pressure drop before and behind the wavefront is huge; reaching the barrier, it has huge destructive power. Such a wave is called the shock or detonation. Detonation combustion mode destroys an engine; it is the cause of terrible explosions of gas-air mixtures (including explosions of household gases in homes). Chemists invented supplement substances which remove and transfer the detonation combustion in normal mode. These substances provide explosion safety of air-gas mixtures, and their wide application has become of great importance in engineering. It is thanks to them that the internal combustion engines function normally, without detonation.

Recall that chemical energy always accompanies us: car batteries, wristwatch batteries, and liquid crystal displays are behind all this chemistry. It plays a key role in the conversion of natural fuels into forms that are available for direct use (production of hydrogen, gasoline, gas from oil, coal, and shale, fuel from household waste and biomass, etc.). In nuclear power, three-quarters of all operations from uranium production to reprocessing of nuclear-spent fuel are purely chemical. The reader can easily multiply this "energy" list of chemistry.

Materials and goods. Any physical idea without chemistry and chemical technology remains empty; it begins to work only being embodied in substances, materials, and constructions. Chemistry is the main creator and supplier of materials. Glass, cement, concrete, asphalt, slate, metal, glue, paint, rubber and rubber, polyethylene and Teflon, fiber, ceramics, semiconductors, and insulators—all are children of chemistry. Polymeric materials for various purposes, for example, engineering materials, having high strength, the superior strength of steel are produced on a huge scale; functional materials are widely used in everyday life (from contact lenses to packaging material). All microelectronics is based on chemicals—silicon, gallium arsenide, etc. Solid state ionic conductors are included in memory devices and displays. Chemical sensors and sensors are used as solid electrolytes and electrodes in batteries and accumulators. An optical fiber one-tenth the thickness of a human hair has replaced the copper wire; thus, the transmission of information through electrical signals is replaced by the transit of optical pulses transmitted through the optical fiber. Modern communications and the Internet are based on fiber optic technologies. There is a struggle for new optical fibers from fluoride glasses; their optical transparency is expected to be so high that they will be able to transmit optical signals and communicate across the Pacific Ocean without intermediate signal amplification at relay stations. And here chemistry is at the forefront.

In chemistry, there is a race for new materials that can work in extreme conditions—heavy-duty ceramics and composites for engines, rocket hulls, nuclear reactors, etc.; heat-resistant materials that can withstand temperatures of tens of thousands of degrees; high-temperature superhard materials for the manufacture of metal-cutting tools, etc. Chemistry and chemical material sciences to a large extent determine the progress of modern technology from washing machines to intercontinental missiles.

Quality of life, culture, art. Washing powders, soap, cleaning agents, deodorants, dyes, coatings and films, finishing materials, skis, roller skates, film, and photographic materials, computer disks, paints and varnishes, disposable tableware, synthetic carpets and clothing, shoes, plastic windows, linoleum, Teflon coated pans—all comes from chemistry. Chemists have developed tests for air and water purity, and sensors that can signal the presence of trace amounts of pollutants. Analytical chemistry and chemical diagnostics guard the ecological safety of the environment. Forensic chemistry has developed sophisticated methods of identifying criminals in almost elusive ways.

Chemistry has penetrated into a culture, into art. We almost do not realize that paper is a product of chemical production; CDs, television screens, paints, photo paper, sneakers, rods made of synthetic materials, Polaroid, decorative coatings, summer ski jumps, glazes and artistic ceramics, colored mosaic glass—there are traces of chemistry in everything. Chemistry is everywhere, even in cosmetics… Remember the joke: the question "What great did chemistry create?" has the answer: "Blondes."

What are on the scales? The objective of chemistry is a multiplication of the necessities of life; this idea was expressed firstly by Russian scientist Mikhail Lomonosov at the seventeenth century. And since then, chemistry has firmly followed this law. Chemistry produces more than good; it creates national and each person wealth. Some discoveries of chemistry have created enormous wealth. For example, the introduction of catalysts in oil refining carried out by the Russian chemist Vladimir Ipatiev in the early twentieth century, a quarter increased the national wealth of the United States (and hence the world). Now the same giant leap is foreseen from genetic chemistry and transgenic engineering. All that was said above about chemistry as a social science is not actually a complete catalog of wealth created by chemistry, including the main ones—health, longevity, and comfort of life.

But chemistry could be dangerous (as nature and civilization: storms and storms, earthquakes and nuclear disasters, explosions and fires, tsunamis and military extremism, etc.). Many of the substances produced by the chemicals are toxic, flammable, explosive. Nature gives no less such dangers (remember at least banal fungal poisons and poisoning), but we are used to them. Yes, chemical plants and warehouses sometimes explode, ignite; but almost always these events are provoked by human negligence, incompetence, thoughtlessness, and sometimes by criminal abuse.

Of course, such explanations do not calm, do not justify, and do not relieve the fear of chemistry and chemical production. You can, of course, give up all the dangerous chemical industries and dangerous things that chemistry produces. But then you have to abandon the gasoline and medicines, from sulfuric acid and batteries, rubber and fertilizers, actually from everything that chemistry produces. Washing powders are almost safe, but the production of benzene with which ones are made is dangerous, etc. There are few people who will agree with such a "refusal" alternative, and the vast majority will rightly consider it demagoguery. Both people and humanity as a whole have learned to find a balance of fears and benefits, to weigh the dangers and losses, risks and comfort.

Gasoline is a very dangerous substance: toxic, flammable, explosive or semi-enclosed volumes, easily evaporates, and the gasoline vapor is capable of detonation. It is produced in huge quantities; hundreds of millions of people carry it with them in the tanks of cars; all countries, cities, and continents are covered with a dense network of gas stations, in the tanks of which millions of tons of this dangerous chemical. However, the benefits and comfort of gasoline cannot outweigh the dangers and fears. In this story, the risk is conceded; humanity deliberately chooses the risk, because the benefits are too great.

An opening of trinitrotoluene and a huge collection of other explosives and propellants based on them had a huge (mostly bad) influence on human life. Explosives are dangerous, it detonates, the whole history of explosives production replete with explosions, fires, disasters. However, no one has put forward a logical proposal to ban it. Here again, the risk wins, humanity deliberately takes the risk, getting along with dangerous explosives. And it is clear why: the benefits of its application are too great in the construction of roads, tunnels, and canals, in the development of ore deposits, etc. And there is a huge military area in which all weapons—from pistol to rocket—act only because there are explosives…

The Beauty and Magic of Molecules

Atoms build a molecule by combining electrical forces; there are no other forces in chemistry. Chemistry recognizes only the Coulomb rule and only the Coulomb potential is her religion… Namely, the Coulomb potentials unite, socialize atomic orbitals (and the electrons) into common molecular orbitals. Electric forces use 16 atomic s-, p-, d-, and f-orbitals as 16 chemical musical notes on which the chemical world sounds. United atoms recognize each other through intramolecular interatomic potentials. These are also electric forces and they control the spatial arrangement of atoms and the bulk form of the molecule. Finally, there are intermolecular interactions, even weaker, but also electrical. They organize molecules into crystals, into films, into clusters, into snowflake stars, into frosty patterns on a glass... Electric forces create the architecture of molecules—delightfully beautiful and aesthetically perfect, simply amazing. They are molecular artists; they create the beauty and elegance of molecules. They can't convey words; this should be seen...

Here is a molecule of adamantane (1a); here the dash is a chemical bond, they connect the carbon atoms. Atoms are at the junction of chemical bonds. The molecule is beautiful by symmetry; it reproduces the crystal lattice of the diamond. By the way, a diamond is one molecule; whatever the size of a diamond crystal is, it will be one molecule, where each carbon atom is connected to four neighbor carbon atoms.

There is a relative of adamantane—Congressene (1b); it was so named because it became the emblem of the International chemical Congress in 1963. Near there is a tetrahedron of carbon atoms (2); the atoms of the tetrahedron can attach any chemical group and create a family of tetrahedrons. Here kuban (3), eight vertices are the carbons with chemical satellites R.

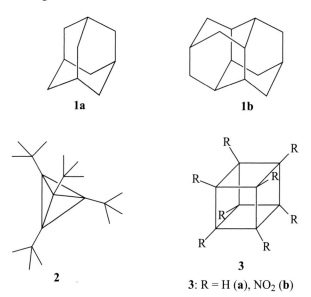

1a   1b

2   3

3: R = H (a), NO$_2$ (b)

Molecules resembling human figures are constructed of carbon atoms. Here, in addition to the single chemical bonds formed by a pair of electrons (one dash), there are triple chemical bonds formed by three pairs of electrons (three dashes). There are also two aromatic hexagons (benzene rings) and a five-membered ring with two oxygen atoms that mimic the head of the figurine. They can be modified by adding new carbon atoms and creating a cap of a jester, professor or baker, as well as a tiara of the Pope or the Royal crown.

Here are two beautiful molecules that are single-molecule magnets. The first one is composed of 12 manganese ions combined into two groups: four central Mn^{4+} ions (pink) and eight Mn^{3+} ions on the surface (red). The molecule includes 12 oxygen atoms, 18 acetate ions (anions of acetic acid), and 20 water molecules. In this molecule, there are 44 electrons, which have a magnetic moment and are tiny magnets (hereinafter we will call them magnetic electrons). Combining, they create a joint magnetic moment, thus a single-molecule magnet is obtained (see below). Another molecule, captivating beauty of symmetry, is composed of eight iron atoms with satellite atoms H, O, and C. It stores 40 magnetic electrons and it is also a molecular magnet.

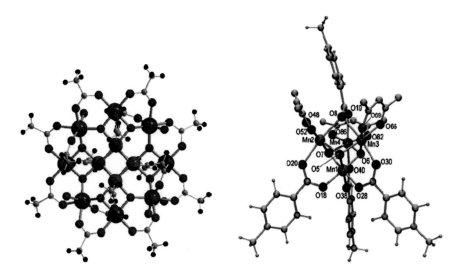

And now—a family of manganese complexes—Mn_4, Mn_8, Mn_{12}, Mn_{30}, and finally Mn_{84} are shown at the next figure. They're all molecular magnets. The latter complex has the form of a closed ring and contains about 300 magnetic electrons.

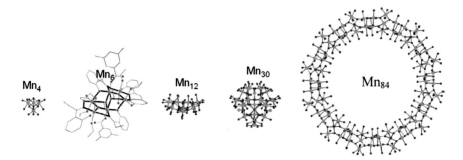

These beautiful molecules fit into even more beautiful crystals. They form cylinders (visible symmetry of the sixth order, the view from the end—see below), which are organized into stacks (side view). See the profile of one of the rings having a thickness of 1.2 nm (one billionth of a meter).

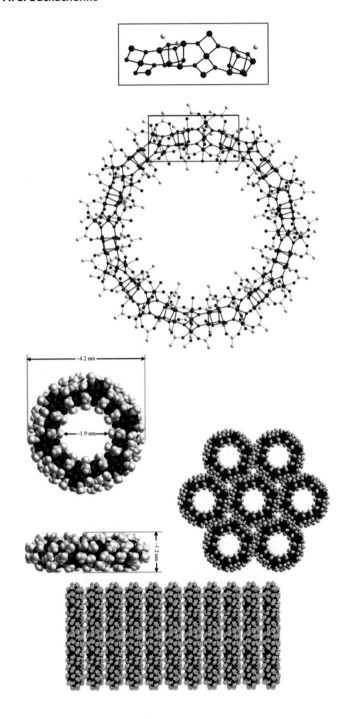

Magnetic molecules are not only beautiful but also useful—behind them there are high technologies in computer science and in the creation of a new generation of memory elements.

But the absolute beauty is the C_{60} fullerene molecule; its charm lies in its symmetry. But the molecule C_{70}, C_{82}, etc. are not so symmetrical—they are stretched into ellipsoids.

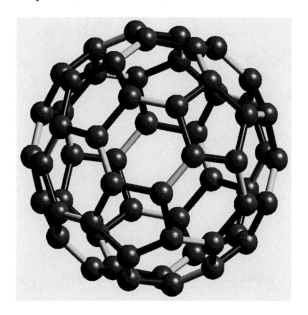

Here are examples of sophisticated beauty—molecular rings, threaded into one another, but not interconnected by chemical bonds. This is catenane; the figure at the bottom of the ring is composed of a chain of rings (on the right they are blue balls for the sake of simplicity and clarity) and carbon atoms.

Below is the image of a molecular machine. Here, a cylindrical molecule (it is depicted in the center as a cylinder) is clad on a molecular rod, at the ends of which the bulk atomic groups $Fe(CN)_5$ lock the cylinder and prevent it from slipping off the rod. The cylinder moves between the ends of the rod, and the motors are electric charges on nitrogen atoms. Charges can be controlled by changing the acidity of the solution (i.e., creating an excess or lack of protons).

The total number of chemical molecules has already surpassed 20 million and about 500 new molecules are added daily. Not all of them are beautiful, but all useful. They are like ladies—not all beautiful, but all are loved (according to a men's logic). Now a lot of substances created by chemists are used. But most are on the waiting list for the moment when they will be in demand.

Life and the Charm of Biological Molecules

All content, all meaning of life is Darwinian evolution. But it is preceded by the chemical evolution of molecules, molecular evolution, evolution at the level of self-reproducing molecules of life as RNA, DNA, enzymes. All of them work as molecular machines, each of them performs its own work, performs its function.

Their lives are governed by interatomic potentials. The potential, i.e., the dependence of energy on distances and angles, is their only Coulomb God.

The birth of these molecules is an unsolved and intriguing mystery. It is a mystery how from very simple molecules—hydrogen, methane, water, carbon dioxide, ammonia, etc.—the DNA molecule was born—a long chain of bases, i.e., nucleotides, connected by the same chemical bonds phosphorus-oxygen. And each chain is connected to another of the same chain in a double helix, and even so that the thymine base of one chain is held by another adenine, and guanine is held by cytosine. You can think of hundreds of ways and dozens of chemical mechanisms for the birth of DNA, but to guess the one that Nature has chosen is almost impossible. A lot of smart people do this, but even more, consider the science of the origin of life a virtual science—because life already exists, and how it arose is not so important. However, this is not a common point of view...

The DNA molecule is beautiful... It is an icon of modern biology, the Mona Lisa of modern science... An outstanding charismatic symbol of our time—the spiral staircase leading to the heavens of knowledge... The keeper of genetic treasures... A huge orchestra of RNA molecules works under her control. This work makes enzymes—molecular machines that produce everything you need for life. Moreover, they chemically control the DNA making mutations.

An ATP synthase is the main energy producer in the cell; both structurally and functionally it is the most beautiful and perfect molecular machine, which reversibly converts the energy of transmembrane potential (it is created by the flow of electrons from the food consumed by the body) into the chemical energy of adenosine triphosphate—the main energy carrier in living organisms.

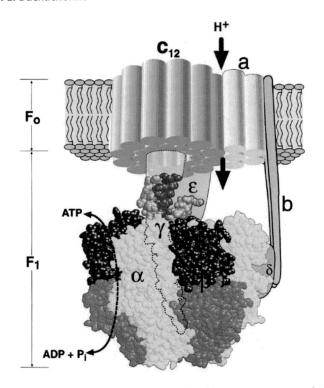

Recall that the ATP synthase is composed of two parts—F_1 and F_0. The first one contains three pairs of protein domains $\alpha_3\beta_3$ and an eccentric shaft γ, which rotates, compressing periodically each pair of domains. Catalytic sites, where both hydrolysis and ATP synthesis occur, are localized in protein β subunits. The second part, F_0, is mounted in the mitochondrial membrane and is the rotor that rotates the shaft γ in the stator $\alpha_3\beta_3$. When ATP hydrolysis and energy release occurs, the shaft γ rotates counterclockwise with the rotor and pumps protons out of the mitochondria, i.e., acts as a proton pump, doing work to increase the proton gradient and transmembrane potential. In contrast, when the ATP synthase acts as a proton turbine, the protons are moving inside the mitochondria and creating a potential difference of opposite sign drives the F_0 rotor and shaft γ in a clockwise direction; this movement is accompanied by the synthesis of ATP in the stator $\alpha_3\beta_3$. It is clear that ATP synthase is a two-stage, reversibly operating an electro-mechano-chemical molecular machine that provides the body with energy.

All life is controlled by molecular machines. The boundary between the inanimate and the living, between the world of the dead and the living molecules—this is the boundary when there were molecular machines. And then they teamed up using interatomic potentials in ensembles, in system, has developed expertise and created the organisms—at first simple and then complex when involved Darwinian, biological evolution.

Biological evolution took the baton of chemical evolution and preserved everything that chemistry created. In fact, all RNA and DNA, all enzymes are almost unchanged transferred from the simplest organisms—prokaryotes—to higher, highly organized—eukaryotes (including humans). The same proteins having 22 amino acids, the same nucleic bases, the same phosphorus-oxygen chemical bonds that form huge polymer molecules of RNA and DNA, the same enzymes with the same functions… Life as a molecular phenomenon is both stable and changeable. Steady in the main and variable in the private… And it is beautiful (especially the one that cannot afford (it is a joke!)).

Science and Art of Chemical Synthesis

Chemical synthesis is the key to all chemistry, the source of all its treasures. This is what makes chemistry the most creative science, which puts it at the center of science. Both science and art are part of the great concepts of culture. But there is a deep difference between them. In art, a unique, inimitable result is a master-piece, there is pride, victory; it brings the author recognition and glory (often eternal and worldwide). In science, a unique, non-reproducible result is a major defeat; it bears witness to an error and brings condemnation and dishonor to the author. The architectural world of molecules, artificially (and skillfully) created by chemistry, is infinitely diverse and fantastically rich, magically beautiful, and often exotic. The rather different molecules were synthesized: rhomboid, poly-metallic chains, polymetallic clusters dressed in organic clothing, fullerene clad metal, crown-molecules trapping metal ions, molecules with exotic topology and architecture. Chemical "vases," cavitand (reversible closed vase), molecular Möbius strip, catenary (passed each other the rings), polycatenanes (chainrings), rotaxanes (rings, dressed on the molecular terminal), etc. are known nowadays.

Much of what chemists create will forever remain exotic but much gives rise to new chemical roads. Because science today's trash can become a trea-sure in the future… Predictions are hopeless here. The great Russian theater director Nikolay Akimov once remarked: "Making predictions is like writing a memoir to a person at birth, and then forcing him to live by it."

A complete synthesis of palytoxin (natural poison, isolated from marine corals), its molecule contains more than 130 atoms. It is the largest and most complex molecule of all, synthesized by chemists, it is a triumph of synthesis as a science and art. This synthesis is an indicator of the power of modern covalent chemistry and evidence that the chemistry of covalent bonds is close to the limit of possible and reasonable. But a new era of non-covalent chem-istry is opening, which uses a wealth of non-valent interactions: hydrogen bonds, complex interactions, Van der Waals forces. This is a huge field in which two giants work together—energy and entropy; the result is the birth of a new molecular chemical architecture that is often superior to what Nature does.

Recall that the covalent chemical bond unites atoms with electrons, which are in their common possession. The negative charges of these binding electrons hold together the positively charged nuclei of atoms. All other, non-valent interactions hold molecules at the expense of the electric charges distributed on them. In passing, we note that entropy is often mystified, idolized, considering it the God of order and chaos. In reality, entropy is a pale reflection of the beautiful and eternal struggle between the electrical potentials that organize order and the warmth that creates chaos…

When Chemistry Cares About the World

There are thousands of events and hundreds of discoveries in chemistry that the world does not know about. Yes, and doesn't want to know… But there are events that make the world shake.

Superconducting ceramics. The discovery of superconducting ceramics shocked the scientific world of chemists, physicists, and engineers. These substances—$YBaCuO_4$ (yttrium, barium, copper, oxygen)—were prepared by chemists (and by Russian chemists long before superconducting properties were discovered). It was such an intriguing and unexpected property that gave rise to unprecedented enthusiasm and hopes for revolutionary transformations in the energy sector. High-temperature superconductivity of ceramics has become a source of delight and great hopes. The Nobel Prize was awarded for this. However, the achievements were much more modest than expectations: the critical temperature of superconductivity stopped somewhere in the region of liquid air and despite the huge efforts of chemists, it stubbornly does not want to move higher. Although progress in this area has almost stopped, there may still be pleasant surprises. This is the case when Niels Bohr was right; he said that problems are more important than solutions: solutions can become outdated, but problems always remain inspiring.

Molecular ferromagnets. Ferromagnets belong to the materials that form the foundation of modern civilization. There are two types of ferromagnets—metal (iron, etc.) and ionic (Fe_3O_4 type) ones. Atoms or metal ions are carriers of magnetism in both types of ferromagnets. However, a new type of molecular magnets has appeared on the chemical scene at the end of the last century. Chemists have been able to produce magnets from organic molecules and to solve a paradoxical problem—to create ferromagnets from organic paramagnetic molecules with antiferromagnetic interaction. The problem is beautiful and it is solved. Molecular ferromagnets will find their place in high technology.

Fullerene. These substances were made from carbon ball-molecules and have become a real treasure of chemistry; here the achievements were above all expectations. They have a high electronic "capacity" and serve as a material for chemical batteries and accumulators for microelectronics. Their salts are superconductors. Fullerenes are the most popular molecules, they went everywhere, even in medicine. Their synthesis is also awarded the Nobel Prize.

Endofullerenes. These are probably the most exotic and wonderful compounds ever created by chemistry; there are atoms, ions, and even molecules captured and trapped inside fullerene spheres. Already synthesized a huge number endofullerenes: $T@C_{60}$ (C_{60} with tritium inside), $He_2@C_{70}$ (two atoms of helium inside the C_{70}), $La_2@C_{80}$ (two captive atom of lanthanum), $Sc_3@C_{82}$ (three atoms of scandium inside); there are buckyballs with the four "prisoners"— atoms of strontium, erbium, and nitrogen: $(Sr_3N)@C_{80}$ and $(ErSr_2N)@C_{80}$.

Carbon nanotubes. A significant event was the discovery of carbon nanotubes—graphite planes rolled into a cylinder. Their diameter is about 100 Å; the ends of the cylinders can be open or closed by "hats," by fullerene hemispheres; cylinder length could be hundreds of nanometers. There is a rapid race in the field of carbon nanotubes in all directions: synthesis, electrochemistry, optics, mechanics, and electrophysics. Nanotubes are chemicals with great potential for use in chemistry (chemical current sources, materials and reinforcing polymers, etc.) and in molecular electronics.

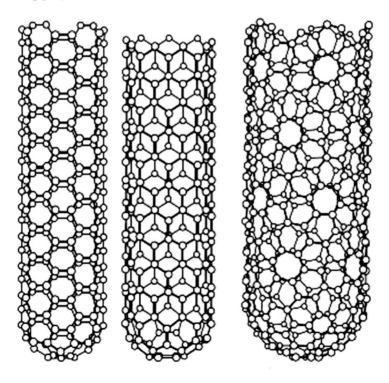

Graphene is a new treasure of the present, born from the garbage of the past… Indeed, graphite has long been known to everyone, its carbon planes are known to schoolchildren. But to isolate and separate a single atomic plane was possible only recently. Graphene is the first two-dimensional material with a thickness of one carbon atom. Each of the carbon atoms carries one electron weakly bound to it. All of them move at a speed that is only 300 times less than the speed of light and therefore graphene has a unique electrical conductivity. Huge expectations, although some smart people think that graphene is waiting for the humble fate of the superconducting ceramic.

Catalysis—a versatile tool to control the chemistry, to direct the reaction along the route, to stimulate or inhibit chemical transformation. A major breakthrough was made in the catalysis of asymmetric synthesis. He opened the way to the synthesis of optically active molecules, and this ensured the success of the chemistry of drugs, physiologically active substances and generally "thin," low-tonnage and high-tech chemistry. But catalysis remains largely an empirical science, a matter of trial and error. Each new discovery becomes a subject of delight, and no one knows when and where it will happen. However, this is true of all science and, perhaps, in such pleasant and happy surprises is the beauty, charm, and magic of science.

Nanochemistry Is a Path to High Technology

Nanochemistry is the science of "nanoworld," the vast world of small particles, the world of small ensembles of atoms and molecules. The main intriguing question of nanochemistry is how the properties of individual molecules are transformed into properties of large bodies, how bridges are built between the world of a single individual molecule and the macroscopic world of matter, how the hierarchy of quantity is transformed into a hierarchy of quality, how the particle size controls their properties. These are the dimensional effects in chemistry. They exist because the number of atoms in small particles is less than the number of potentials and therefore they have free vacant potentials. They work in nanochemistry.

Nanochemistry has already created wonderful things: the molecular diode and transistor, ferromagnetic and semiconductor nanowires, electronic shuttle nano-pendulum, atomic-molecular pump, and the nano-laser. And all this is an element base of new molecular electronics, the prologue to a new technological civilization.

Nano-scale technologies have opened the way to the creation of microscopic mechanical devices—machines, robots, motors (with the prefix nano).

Here's an example of a chemical nano-engine. By electrolysis, a metal nano-rod the size of a bacterium is created (length 2 microns, i.e., 2 millionths of a meter, thickness 350 nm, 350×10^{-9} m), one end of which is made of gold, the other—of platinum. The rods are placed in an aqueous solution of hydrogen peroxide, in which they begin to move, moving like torpedoes along a long axis. The driving force is created by chemical reactions of hydrogen peroxide. The mechanism creating the force is as follows. The peroxide molecule decomposes at the platinum end of the rod (platinum catalyst):

$$H_2O_2 \rightarrow O_2 + 2H^+ + 2e,$$

giving birth to a molecule of oxygen, two protons (they remain in water) and two electrons (they go into the metal). At the gold end of the rod is the reverse reaction:

$$H_2O_2 + 2e + 2H^+ \rightarrow 2H_2O,$$

where the electrons are transferred from the metal to the molecules of hydrogen peroxide, H_2O_2, forming two HO^- that take two protons from the solution, giving two molecules of water. The total result of the reaction is as follows: on the gold end protons are consumed, and on the platinum end they are formed. A proton concentration gradient is created. But protons are strongly bound to water and therefore the movement of protons along the concentration gradient (from platinum to gold) means the hydrodynamic movement of the surface water layer from platinum to gold. According to the law of conservation of momentum, the rod must move in the opposite direction. And that's what really happens: the nano-engine is running. But it is possible to fix the rod and then it is converted into a nano-pump, which moves the liquid, creating its flow.

Already the smallest radio receiver is built and running in which the receiver of radio waves and the emitter of sound waves is a single carbon nanotube. It is made so. A nanotube with a length of 500 nm (50 millionths of a meter) and a diameter of 10 nm (about the size of a virus) is fixed on the electrode or grown directly on it by the method of deposition from vapors. Opposite the end of the tube, another electrode is installed and a potential difference is created between them. An electric current appears. Radio waves modulate this current and create mechanical vibrations of the nanotube with the same radio frequency. That is, the tube works as an antenna: it is tuned to the radio signal, amplifies it, converts it to sound and transmits it to the loudspeaker. This is a completed project: you can even listen to the song on the website www.SciAm.com/nanoradio.

Nano-radios promise a great future in the miniaturization of devices such as hearing AIDS or cell phones. There are ideas to use such a receiver as a nanosensor: the impurity molecules adsorbed from the atmosphere (for example, organomercury compounds) change the mass of the nanotube and the resonant frequencies of its own oscillations. These changes can be detected by the frequency shift of the received radio signal. And it will be a detector of unsurpassed sensitivity; a dozen molecules are enough to detect them.

The third example is a molecular transistor on a fullerene molecule. A fullerene ball with a diameter of 7 Å (seven-tenths of one billionth of a meter) is placed between two metal electrodes on a silicon dioxide substrate (this is well-known sand). When a small electric potential (enough thousandths of volts) is applied to the electrodes, an electron flow is established between them. They jump from one electrode to a ball of fullerene and then jump on to another electrode. At the moments when the electron is on fullerene, it creates a charge on it and then the molecule is repelled in turn from one or the other electrode. Coulomb repulsion causes the molecule to jump from one electrode to another at a frequency of about a trillion (10^{12}) jumps per second. This is the first single-molecule transistor in the history of civilization. And it is quantum, because the current in it does not change continuously, as in the classical transistor, but jumps discretely.

Chemical Aroma of Physical Superconductivity

Aromaticity is a fundamental property of chemical particles to collectivize weakly bound electrons, to make them common and to form an independent electronic subsystem. Most of the electrons in the molecules are held firmly by their host atoms; they occupy the internal atomic orbitals near the nucleus. However, the electrons have no position in the inner orbit, they are in the outer orbit, and they are held very weak by the nucleus. They are united in an autonomous system and become delocalized, i.e., pass into the collective ownership of all atoms. They do not occupy atomic strictly localized orbitals but belong to molecular delocalized orbitals.

A physical characteristic of aromaticity is closed circular electric currents generated by the sharing of electrons. Such currents can be created in flat structures (e.g., graphite and graphene), or spherical (type C_{60} fullerene), or cylindrical (carbon nanotubes). Each carbon atom in such molecules has six electrons. The atom retains four of them and gives two ones in common use; it is the atomic p-electrons (their orbital angular momentum is equal to one), which in the molecular orbitals are renamed π-electrons.

Atomic and molecular electrons form an electronic topography on the molecular surface; thus, in two-dimensional graphite, this relief is strictly periodic—antinode of the electron density on atoms and nodes it in the interatomic regions (the alternation of hills and ravines). Circular closed currents in aromatic molecules create an orbital angular momentum and the corresponding magnetic moment and magnetic field. When aromatic molecules are placed in an external magnetic field (for example, when structures are analyzed using nuclear magnetic resonance, NMR), the magnetic field of the circular current partially shields the external magnetic field of the NMR spectrometer. As a result, the NMR lines of aromatic molecules are strongly displaced across the field, and this shift is a quantitative measure of the aromatic molecules.

What is the relationship between the two fundamental properties—aromaticity and superconductivity? Both are because the electrons are "unbound". There is also a difference: aromaticity is considered to be a property of common electrons at the scale of one molecule, and superconductivity implies large-scale collectivization at the intermolecular level. Aromaticity is a property of molecular superconductivity; this property is supramolecular and macroscopic. Graphite as an aromatic structure does not exhibit superconductivity (it may lie in the temperature region, not reached experimentally yet). However, a single carbon nanotube, i.e., a graphite atomic plane rolled into a cylinder, is a real superconductor with a transition temperature to a superconducting state depending on the magnetic field (such dependence is a sign of superconductivity). Of course, the aromaticity of chemical structures does not have to be accompanied by their superconductivity, but a broad view of these two phenomena can suggest new ways to superconducting structures.

Physical Mirror of Chemistry

The great Feynman regarded chemistry as the most complex physics; the reason is simple—he knew it poorly. The physics of chemical reactions is the elite part of chemistry, the bridge from physics to chemistry, their common territory... In the physics of chemical reactions, there are two parts—statics and dynamics. The first is quantum chemistry, which has mastered the basic principles of quantum mechanics and adapted them to solving chemical problems: structure and stability, thermodynamics and energy, geometry and physical properties of chemical particles and substances. The most beautiful is the design of potential energy surfaces for multi-electron and multi-atom reacting systems. An example of such a surface for a simple reaction $Br + I_2 = BrI + I$ is shown below.

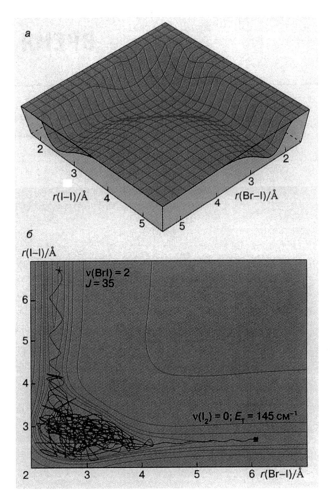

The potential energy surface is at the top of this Figure (a), where reagents (bromine atom Br and iodine molecule I₂) are in the left valley, and products (iodine atom I and bromine molecule Br₂) after the reaction will be in the right valley. At the bottom (b) there is a top view; it shows how intricate and painful is the trajectory of reacting particles moving from the valley of reagents to the valley of products.

The theory of the chemical transformation of reagents into products describes the movement of the reacting system from one valley to another with the overcoming of the barrier separating these valleys. It analyzes the factors controlling this motion (energy of different molecular states, orbital symmetry, angular momentum, electron, and nuclear spin, velocities of atoms and molecules, etc.).

The main concern of quantum chemistry is the solution of the Schrödinger equation.

$$H(r, R)\psi(r, R) = E(r, R)\psi(r, R).$$

For those who know: the solution yields the molecular wave function $\psi(r,R)$, which depends on $3n$ coordinates of all n electrons, denoted collectively as r and nuclear coordinates collectively denoted by R. It also gives the surface potential energy $E(r,R)$ as a function of the nuclear coordinates R. This equation cannot be solved exactly, but you can reduce the wave function of $3n$ variables to n functions (molecular orbitals), each depends on three variables (coordinates). Such a one-electron molecular orbital (MO) describes the distribution of one electron that moves in the average Coulomb field of all other $(n - 1)$ electrons. This is the first fundamental approximation that gave rise to the rapid progress of quantum chemistry. The well-known Hartree–Fock method based on this approximation is simply the mean field method applied to the multi-electron problem, and for this reason, it is called the self-consistent field method.

The second key approximation of quantum chemistry is the decomposition of unknown MO functions into known atomic wave functions (atomic orbitals, AO), which describe the distribution of electrons centered on an atom. AO follows from the solution of the Schrödinger equation for the atoms that make up the molecule; they are almost exact. These hydrogen-like AO (or Slater orbitals) are usually taken as linear combinations of Gaussian functions; such AO decomposition greatly simplifies analytical calculations of integrals when solving the Schrödinger equation.

If AO and their Gaussian decomposition are chosen intelligently and if their number (base series) is increased, more and more accurate MO are obtained. So constructing a complete and accurate basis series is a constant concern of quantum chemistry. There are a number of methods for achieving a high degree of completeness of the basic functions; in fact, they have become ready-made recipes and exist as ready-made computer programs.

In 1929, Paul Dirac optimistically stated that "the fundamental physical laws necessary for chemical theory as a whole are fully known"; later Lewis warned that "the problem of many bodies (electrons and nuclei) that make up an atom and a molecule cannot be solved to the end." This foresight of genius identified the main problem of quantum chemistry from Lewis to the present day.

The Hartree–Fock theory of the self-consistent field describes the major part (about 99%) of the total energy of the molecule. But it is the remaining part (~1%) that is critical for the chemical bond; it arises because of the forced neglect of the instantaneous interaction (correlation) between electrons. This correlation energy (or electron correlation energy) is the difference between the exact energy of the molecule and the self-consistent field theory. It is unobservable; it is not perturbation energy that can be turned on or off. This is an inevitable error of the self-consistent field method, arising from the fact that the exact wave function of the molecule is composed of atomic orbitals. But without this representation, the Schrödinger equation is generally useless; one can say that the correlation energy is a victim of quantity in favor of quality, that it is an inevitable, forced artifact. And the whole history of quantum chemistry is a grueling struggle with this artifact, a struggle for methods that could calculate this mythical correlation energy as accurately as possible that could compensate for the weaknesses of the self-consistent field method.

In the last decade, a new method alternative to the classical self-consistent field (SSF) has been developed; it was called density functional theory (DFT). DFT works with the electron density distribution $n(r)$, not with the multi-electron wave function $\psi(r,R)$. Since the electron density $n(r)$ is a function of only three variables (unlike $\psi(r,R)$, which depends on $3n$ variables), DFT greatly simplifies quantum calculations. The theory calculates the energy as a function of only the electron density and the fixed nuclear coordinates. Moreover, there is no problem of electronic correlation at all, since DFT automatically includes ready-made "correlated" electron density.

Now a few notes should be a mention about the "dynamic" part of physics, about the theory of chemical reactions. Its task is the development of methods for the calculation of the reaction rates for given quantum states of the reactants. Any calculations start from the potential energy surface (PES), which is the conductor of the chemical reaction; it controls the trajectories of transitions from reagents to products, determines the reaction channels and the energy distribution in the products. There are many semi-empirical and non-empirical methods for calculating PES; they are based on quantum chemistry methods. Classical and quasi-classical methods work quite accurately, especially for large masses and energies. Although chemistry is a quantum science, almost 99% of all chemical processes are described satisfactorily by the equations of classical mechanics with quantum corrections. And it is again a great success (or failure?) for chemistry: it is just that the world is such that chemistry, in its energy parameters, falls into the boundary area between classical and quantum mechanics; that is why most of it is described by the laws of classical Newtonian motion, but with amendments to the "quantum" of motion. Strict, purely quantum solutions of the Schrödinger equation are required only in the field of low energies and low masses (usually extreme or exotic conditions: ultra-low temperatures, etc.); but even for these cases, a semiclassical version of the theory has been developed that allows quantum corrections to be included in classical chemical dynamics. Physics of chemical reactions is a beautiful area where classical and quantum physics work together.

Chemical Physics as an Atomic and Molecular Culture

There are a dozen scientific institutes of chemical physics in the world. However, chemical physics is more than science. It is a great scientific culture in which the world is known as the unity of physics, chemistry, mathematics, and biology. It is a culture in which thinking and knowing do not recognize professional boundaries…

Rich Life of a Single Molecule

We have already talked about the outer beauty of molecules; now we will talk about the inner beauty, the beauty of their lives… Chemistry has reached the upper limit horizon—the ability to see a single molecule, to measure almost all its properties, to observe its chemical transformations and functioning. Chemistry had reached such perfection when its main God—molecule—has acquired individuality, has become the person. And it happened only 70 years after the great scientist and philosopher Ernst Mach authoritatively stated that atoms and molecules are not observed and all talk about them is meaningless. Now everything became available to measurement on one molecule. This is the discovery of a new world in the chemistry, of the world of individual molecules, not bound by "ties of collectivism."

And this is a great breakthrough in chemistry, physics, and mechanics. First, the detection of a single molecule puts an analytical limit; single-molecule detectors and sensors of higher sensitivity can be created at this limit. Secondly, for the first time, it is possible to establish individual "personal" properties of the molecule, not averaged and not "hidden" in a large ensemble like it. Third, the life of single molecules became available. It is possible to observe how it works—regularly or randomly, whether to keep her memory about previous events in her life, whether to "rest." In fact, this is a new, beautiful area of molecular knowledge.

Finally, developing technologies for manipulating single molecules, the researchers gain experience in creating elements of nano-electronics, nanomechanics, and nano-optics, which will be an advanced technology in the future. And it is not a declaration, it is a real work on the way of searches and finds, on the way to molecular electronics and single-molecule chemistry.

Luminescence of Single Molecules

It is simple to observe single molecules: laser irradiation of a certain area in which the luminescent molecules are located excites their glow and then they are observed as molecular "fireflies." Their images look as shown in the picture below.

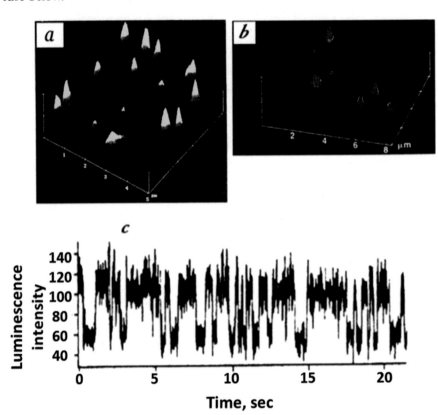

It shows the glow of single molecules—rhodamine (a), cholesterol-oxidase (b), and flicker of one of the molecules of cholesterol-oxidase in time (c). (Numbers of emitted photons per unit time are on the ordinate axis.)

Having lit up, a single molecule has become a molecular sensor, reporter, informant, witness to all the events in the place it occupies—in solids and liquids, on surfaces and interfacial boundaries, in films, in liquid crystals and gels, emulsions and porous media, membranes, proteins, nucleic acids, enzymes, chromosomes, and molecular biological motors. Exciting constant irradiation and observing in time the luminescence of molecular "fireflies," you can observe the movement of molecules on the surfaces and in the volume, to see the trajectory of their movements.

Chemical Life of a Single Molecule

How does the chemical plant, which consists of one employee—one molecule—work? How does one molecule live and function? Here is an example: the figure shows the recording of the account of photons emitted by the reaction product of one molecule—oxyreductase; it is an enzyme, which decomposes hydrogen peroxide and produces a luminous molecule phosphor. There are three records from three molecules 1, 2, and 3.

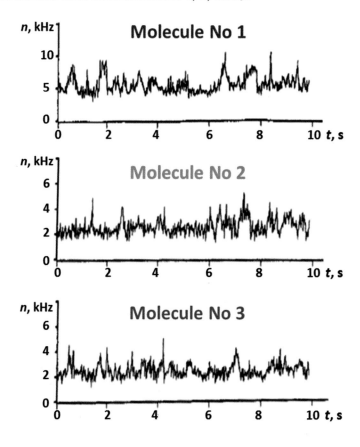

It can be seen that the periods of intensive work of the enzyme alternate with the periods of its "silence," rest. In this case, all three enzyme molecules "flicker" differently during observation (10 s). This remarkable technique is widely included in molecular biology.

Magnet in One Molecule

A single molecule, which carries tens and often hundreds of magnetic electrons, is a single-molecular magnet and demonstrates all its properties: spontaneous magnetization, anisotropy, and magnetization hysteresis, domain organization, dynamic hysteresis. Electrons and their magnetic moments usually "sit" on the atoms of metals concentrated in the center of the molecule. They make up its magnetic core, surrounded by a non-magnetic shell, which serves as "clothes" of organic molecules that make the core magnetically isolated. Single-molecular magnets are synthesized by classical methods of chemistry, isolated and purified as individual chemicals.

The magnetic electrons form into groups and form domains inside the molecule. Unlike classical ferromagnets in magnetic molecules, domains change the direction of magnetization by quantum mechanism, so a quantum hysteresis loop appears in these molecules; its characteristic feature is "steps" of magnetization (figure below). Here M is the magnetization, H—magnetic field strength in kilooersted.

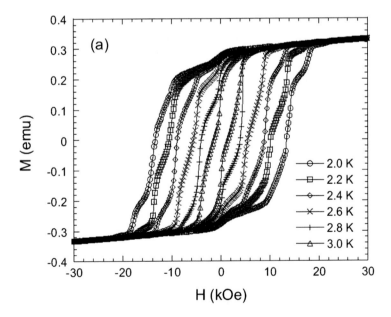

Magnet molecules attract great attention as elements of high-density magnetic memory, as elements of quantum computers; they have a large magnetic moment and are reliably detected and controlled by "point" methods of atomic force and tunnel spectroscopy (see below).

The main fate of these molecules lies ahead when future talented chemists and physicists come up with such structures that could remain magnetic at room temperature. So far, only such structures are known that retain magnetism only at a temperature of liquid helium. The language of magnetic molecules suggests encyclopedic education of their creators…

Portrait of a Molecule: Inside and Outside View

Physicists came up with a great technology tunnel microscopy and created the device—tunnel microscope. It works like this: a thin needle is installed above the surface and keep it away from the surface by atomic distance. Measure the electric potential between the needle and the surface (quite a few millivolts). If there is a negative potential on the needle, a rain of electrons pours from it to the surface and their flow is measured as an electric current (its value at the level of nanoamps, i.e., about 10^{-9} Å). The current depends on the removal of the needle from the surface. Now, with the help of electronic control systems, the needle is forced to slide over the surface. And since it is uneven, the electric current instantly tracks all its irregularities: he sees its mountains and valleys, gorges and ravines, peaks and pits. And all this on an atomic scale! The needle produces "topographical" monitoring of atomic and molecular surface topography. It looks at atoms and molecules of this relief, it gives their images, their external portrait, and it does this as a satellite looking at the Earth from space.

But the sliding needle floating above the surface can be stopped and made to stand over a single atom or molecule for any length of time. Now you need to slowly change the potential of the needle. Then the energy of the electrons emitted by the needle changes smoothly as far as they are accelerated by this potential. And at the moment when the energy of the electrons emitted by the needle coincides with the energy of vibrations of any chemical bond in the molecule (this is resonance!), there is a current surge (due to the resonance scattering of electrons). Each chemical bond corresponds to a current surge at a certain characteristic potential. Electrons shine through the molecule as an x-ray in the doctor's office shines through a person. Each jump in the tunnel current "marks" the vibrational frequency, and their totality is the vibrational spectrum of the molecule. So the frequencies of molecular vibrations are measured, so an inner spectroscopic portrait of a molecule is written (by electrons rather than oil paint!).

Stopping the potential of the needle on the selected oscillation, it is possible to carry out the energy pumping of this oscillation and stimulate the chemical transformation of the molecule, selective for the selected chemical bond. This can be done by bringing the needle closer to the molecule, i.e., increasing the tunneling current. It is actually a new version of selective photochemistry; but it is photochemistry without light, it is "dark" photochemistry. The role of light quanta is played by tunneling electrons, selectively activating a given chemical bond.

In the mode of tunnel spectroscopy, you can not only see a portrait of a molecule (both outside and inside), but also move it with a needle in the mode of "pushing" or "dragging," combine atoms and molecules in an ensemble of a given design, write and draw atoms as paint... Thus it is a prologue to a new technological civilization—to molecular electronics, which works with voltages in millivolts and currents in milliamps. Now it is a well-mastered science that has become molecular engineering. Magnetic resonance of a single electron and nucleus, electrical conductivity, mechanics and mechanochemistry of single molecules and macromolecules have become available—all these elegant technologies are well mastered and work perfectly. And there are areas where single-molecule technologies bring fundamentally new information and a new level of understanding, these areas are a "biological wing" of chemistry.

Mechanics of a Single Molecule

To study the mechanics of a single molecule, atomic force microscopy has been created. It is similar to tunneling but measures the force of attraction between the microscope needle and the surface (as opposed to tunneling microscopy, which measures the current between the needle and the surface). In the classic version, atomic force microscopy is non-contact: the needle slides along the surface without touching it and draws the force relief of the surface: its folds, pits, mountains and all that lies on it. And also on an atomic scale. This is its great value.

But the price of atomic force microscopy increases many times if it is made contact: to fix one end of the molecule on the needle and the other on the surface. Then, pushing the needle from the surface and stretching the molecule, you can measure the force as a function of the stretching of the molecule, i.e., to analyze the mechanics of single molecules. This idea was brilliantly implemented experimentally; the main intrigue was to fix the ends of the molecule (usual macromolecule) by covalent or coordination bonds. This is achieved similar to fishing and multiple diving needles are used as a fishing rod. The molecule is extracted from the surface like a fish from a river.

The figure below shows the scheme of the experiment and the schematic dependence of the force F on the stretching of the molecule x. The force F is usually not linear: it has small local peaks; it means that F rises and decreases. This is evidence that the extension of the folded macromolecule is its unfolding and stretching. Finally, with some critical elongation, the completely elongated molecule breaks and the braking force is the strength of the chemical bond.

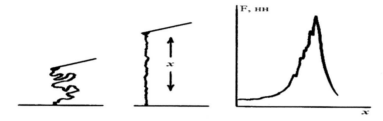

This remarkable technology is used to measure the mechanical forces of molecular motors and biomolecules (enzymes, proteins, nucleic acids, antigen-antibody complexes, etc.). It opens up almost unlimited prospects in biochemistry and biophysics. Although this technique is laborious and delicate, aristocratic, the results are so attractive that its future is brilliant, especially in molecular biology.

Chemical Parade

There are 10^{21} molecules in a drop of water. All these trillions move continuously, arbitrarily, chaotically, randomly; each molecule has its own, independent, individual destiny. Also, the molecule reacts randomly, if it is given the chance to do it. Is it possible to make these trillions of molecules move, rotate, oscillate and even react in concert, synchronously, simultaneously? That is to build them as cadets at the military parade: all cadets as one, all molecules as one... (Physicists call such behavior coherent.)

Of course, such forced "collectivization" of molecules is possible. It opens a new face of chemistry—its property to form oscillatory reaction modes. Coherence, i.e., the synchronicity of the reaction in time, is manifested in the periodicity of the reaction and is detected as oscillations in the output of products, in the flicker of luminescence, electrochemical current or potential, etc.

Chemical coherence exists at two levels, at quantum and macroscopic ones. In the first case, the reactivity of reacting particles ensemble is prepared instantly (for example, by a laser pulse) and behaves coherently. Such an ensemble oscillates between states either capable of reacting, or incapable. Thus, the concentrations of active reagents change periodically in time, and these oscillations modulate the output of the reaction products. Quantum properties are the origin of vibrational and electronic coherence; vivid examples of macroscopic coherence are a synthesis of adenosine triphosphate in living organisms, which provides the rhythmic muscle contraction that runs the athlete and strikes his heart. The average person's life is three billion heartbeats, and all these are coherent chemistry... And the most sensitive and significant in vivo are thinking, remembering, generating ideas and all these are also coherent chemistry.

Coherence introduces new concepts into classical chemistry: wave packet, phase, loss of coherence (decoherence), interference, bifurcations, phase portrait, strange attractor, and phase turbulence. And it's not just a new language for chemistry; it's a new level of thinking, a new technology for chemical research. Incoherent chemistry, random, statistical behavior of molecules is replaced by organized, ordered, synchronous behavior: chaos becomes order.

Quantum Coherence

The period of oscillation of atoms in the molecule is 10^{-13}–10^{-14} s (100–10 fs; 1 fs is equal to 10^{-15} s). To create a synchronous, coherent motion of atoms in the molecule and monitor its behavior over time, it is necessary to act on the molecule very quickly, for times shorter than the time of one oscillation of atoms. Modern lasers generate short pulses of the excitation light, lasting 5–10 fs. The development of such lasers has generated a new field of chemistry called femtochemistry. The ensemble created by the laser pulse is a coherent wave packet that is an ensemble of oscillator molecules with a fixed phase of oscillations, a given starting interatomic distance, and given energy. Such behavior is absolutely synchronous and consistent; so a regiment of soldiers moves on parade.

Electronic movement is much faster than oscillatory; therefore, to prepare an electronic wave packet (an ensemble of molecules in which electrons move in a coherent manner), it is necessary to excite the movement of electrons by even shorter laser pulses, with a duration of about 1 as (10^{-18} s). Single attosecond pulse produces an electronic wave packet, in which it is possible, for example, to create a forced circulation of π-electrons in the benzene ring. The creation of electronic wave packets also implies the control of their dynamics. For example, in electron-coherent ensemble of molecules HD^+ prepared by short ionizing laser pulse, single electron oscillates between the nuclei of hydrogen H and deuterium D. Further, the second pulse is able to exercise coherent control of the dissociation reaction HD^+, for example, to induce the decay of HD^+, either into $H + D^+$ or into $H^+ + D$, depending on at what point the second pulse is turned on. The first channel is possible when the electron "sits" on hydrogen and the second one when it "sits" on deuterium. This is the highest level of coherence. And not only in chemistry…

Coherence in a Heart

The first observation of oscillating chemical reactions has become a part of history; then, the oscillation was perceived as rather exotic rather than physical regularity. The realization that macroscopic coherence is a fundamental property came recently and stimulated interest in chemical oscillators. The most intriguing is the behavior of systems of chemical oscillators linked chemically (two or more oscillating reactions in one vessel) or physically (oscillators in different vessels, but linked by the exchange of reagents due to diffusion or mass transfer, or by the exchange of electrochemical currents or potentials).

In the simplest system of two oscillators, there are three main modes:

- Death of oscillators when they mutually destroy their own oscillations
- Random irregular breakdowns between order and chaos
- Synthesis of the new oscillators

And if the reader condescendingly thinks that these modes are boring, far from his interests and do not touch him, then he is mistaken: the coherence everybody carries within ourselves. A healthy heart is an ideal coherent system; the rhythmic contractions of this organ are the periodic and coherent propagation of chemical waves, the result of the ordered interaction of a huge number of chemical oscillators. The figure below depicts a wave of electrical potential in the heart; the movement of such a wave is the rhythmic coherent work of the heart, this tireless worker.

Three cardiac pathologies are a heart attack, fibrillation, and tachycardia; those are exact matches to the three modes of the simplest system of two oscillators, which were discussed above.

Molecular Cinema

The central event in the chemical world is the chemical reaction, the birth of a molecule, i.e., the rearrangement of atoms and the transformation of their electronic "clothes." The duration of this event is $10^{-11}-10^{-13}$ s. Is it possible to see a live transformation of the reagent molecules into the product molecules, i.e., to observe the movement of atoms at the time of the reaction?

To shoot such molecular cinema, it is necessary to do each frame for a short time 10^{-12} s, i.e., to shoot a trillion frames per second! Recall that the usual movie is 24 frames per second. Feel the difference… However, such a superfast movie has already done. Moreover, the molecular movie was generated, which was called femtochemistry. It's chemistry at times 1–10 fs, it's a brilliant breakthrough in chemistry. The American physicist Ahmed Zewail, awarded in 1999 by the Nobel Prize, was the main mastermind. Femtochemistry investigates ultrafast reactions such as the movement of the electron between the protons in the molecule H_2^+ (~2 fs), the migration of π-electrons in the benzene molecule, and the circulation of the electron around the atom. And of course, femtochemistry is needed in molecular biology; it is already beginning of its rapid invasion into the chemistry of the living, and new breakthroughs in both sciences should be expected.

Chemical Tyranny

Chemistry is a kingdom, which is reigned by Queen, that is the energy, and by King, that is the angular momentum, spin of reagents; they allow or prohibit reaction. The Queen rules graciously: if the energy is not enough to overcome the energy barrier of the reaction, there is always a "workaround"— seeping through the barrier. It's called tunneling, and it exists because there is quantum mechanics that allows atoms to be a wave. And the wave does not recognize barriers. Queen looks these nuclear mischief condescending. The King is cruel and categorical: only chemical reactions that do not require a spin change are allowed; if the spin of the reagents is not equal to the spin of the products, the reaction is strictly prohibited. And no electric power can change the spin; that is the power of the King's dictatorship.

Magnetic Chemistry and Chemical Radio-Physics

Any dictator, including a chemical one, can and should be deceived—he deserves it… The fact is that electrons and nuclei, being quantum whirligigs, acquire magnetism and become magnets. So magnetic forces and magnetic interactions come into play. Being negligibly small—they are millions of times weaker in energy than electric interactions—magnetic interactions change the spin of reagents and switch reactions between forbidden and permitted channels. Thus they remove the spin ban, destroying what was erected by the second dictator. So they write their own magnetic scenario for the new "magnetic" performance, that was put by the chief director—a chemical reaction.

There are new wonderful phenomena and events in the scenes of this play. You can control the direction and speed of the chemical (and therefore biochemical!) reactions by an external magnetic field. It is possible to orient and align the electrons and nuclei spins (similar to elementary magnets) in the molecules produced in the reactions. Physicists call it the creation of targets with polarized electrons and nuclei, but they do it by their own, physical methods. In molecules born in chemical reactions, a huge polarization of nuclei is achieved, often exceeding what physicists obtain. If nuclear magnets are all aligned and oriented against the external magnetic field of the Earth (it is called a negative polarization, chemists can do it easily), such molecules store energy and are able to give it in the form of radio or microwave radiation.

Here is an example of 100 MHz radio emission in the photochemical reaction of porphyrin with quinone. The first arrow indicates the moment the light is switched on, i.e., the start of the photochemical reaction. After 5–6 s, radio emission of negatively oriented protons of quinone molecules appears which participate in reversible reactions, acquiring a nuclear orientation. Radio emission can last indefinitely—as long as there is a reaction that produces a nuclear polarization pump. The reaction stopped (the second arrow indicates the moment of turning off the light)—the pump ended, the radiation disappeared…

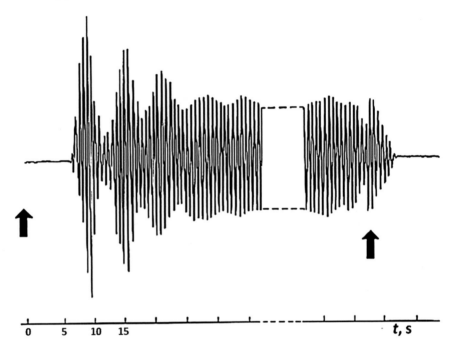

This is how a chemical-pumped quantum generator works; in fact, it is a chemical radio station where each of the reaction products emits its own frequency and has its own voice. And it is not necessary for this photochemical pumping, radio emission was observed in conventional thermal reactions… This is the start of a new delightful area of magnetic chemistry—chemical radiophysics…

Magnetic Isotope Effect and Magnetic Isotope

A new area of chemical science—spin chemistry—was born to deceive the abovementioned king. It deals with angular moments (spins) of electrons and nuclei and develops new principles of reaction control using magnetic fields and magnetic nuclei. The most striking phenomenon in spin chemistry is the magnetic isotope effect. Its meaning is that the rate of chemical reaction depends on the magnetic moment of the nuclei of reacting molecules. The reason for it is as follows. If there is a magnetic nucleus in the molecule, its magnetic field changes the spin and removes the spin ban: the reaction is allowed… If the nucleus is non-magnetic, then the chemical reaction is forbidden.

The difference in the rates of chemical reactions of molecules with magnetic and non-magnetic isotope nuclei is a direct way of separation, fractionation of magnetic and non-magnetic isotopes. This is the basis of a new magnetic isotope effect. Unlike the old, classical and well-deserved mass-dependent isotope effect, which fractionates light and heavy nuclei, the new isotope effect separates magnetic and non-magnetic nuclei.

It does even more: the presence of the magnetic nucleus, removing the spin ban, opens up additional channels of reaction, increasing its performance. Thus, the synthesis of adenosine triphosphate (ATP), which is the main energy carrier in living organisms, is produced by enzymes. They all work in the presence of magnesium ions. Magnesium has three isotopes: ^{24}Mg, ^{26}Mg, and ^{25}Mg. The first two are non-magnetic, the third is magnetic; its nucleus as an extra neutron added to the ^{24}Mg nucleus. If the enzymes contain the ^{25}Mg isotope instead of ^{24}Mg ions, they produce twice as much adenosine triphosphate. If ^{25}Mg ions are delivered to the heart (such target delivery methods have already been created), then it is possible to stimulate strongly the synthesis of adenosine triphosphate in the heart muscle and create a new nuclear magnetic cure for heart diseases arising from a lack of ATP. This is an amazing thing—one extra neutron and wonderful medicine! And it's tested and works on rats, rabbits, and goats. This is just the beginning. A new nuclear magnetic isotope effect promises major discoveries in physics, chemistry, biochemistry…

Magnetic Isotope in Geochemistry and Space Chemistry

The triumphant march of isotopes in science and in life began from the moment of their discovery. Their majestic role in chemistry, biochemistry, geochemistry, historical chronology, art, archeology, ecology, energy, and medicine is known. Isotopes are memory carriers of the birth and transformation of molecules; the distribution of isotopes is the chemical history of a substance, its "chemical Bible." The science of isotopes and their use has become part of culture and civilization.

Isotopes serve as a memory in two functions. First, they participate in the creation of memory and in storage through isotopic effects in the acts of birth and transformation of molecules. Secondly, they are the heirs and keepers of memory; they act as labels and as witnesses of chemical events—both present and ancient ones, which took place for many millennia before our days. Isotopes as witnesses, as heirs, and keepers of chemical memory are actively exploited in ecology, geology, organic geochemistry and biogeochemistry, paleontology, archeology. Suffice it to recall the radiocarbon dating method that used radioactive isotope ^{14}C. It is also appropriate to recall the romantic tale of the Turin shroud, which allegedly harbored the body of Christ after removing it from the Cross...

The triumphal march of isotopes continues... The magnetic isotope effect is a fundamental property of Nature. It controls reactivity through magnetic interactions and is accompanied by fractionation of magnetic and non-magnetic isotopes. The distribution of isotopes created by this effect is the nuclear spin memory of the birth of the molecule, the long-term memory, the term of which is set by the lifetime of the molecule itself. For this reason, the magnetic isotope effect is universal: it preserves the memory of the biochemical, geochemical, and space chemical origin of molecules. This is the memory of the chemical evolution of matter as a combination of a huge number of chemical reactions, which sorted isotopic nuclei by mass and magnetic moments for millions of years. They brought to us the echo of this sort, through which it is possible to reconstruct the pathways of chemical evolution, the origin, and fate of substances in Nature (including ores, minerals, oil, coal deposits, etc.). They left the nuclear magnetic seal of the history of the Earth and the Universe, the seal of the chemical evolution of matter. Perhaps they are able to answer the question of whether there were chemical processes in different parts of the Universe. After all, if there is no chemistry, it makes no sense to look for biology...

Isotopes and Climate

Isotope chemistry works not only for chemistry and biology but also for geology, to recreate the geological history and history of climate. Thus, it is known that the oxygen isotope ^{18}O is distributed unevenly between ice and water: ice is depleted by this heavy isotope, and water is enriched. In the cold periods of Earth's history, ice is abundant and then water contains a lot of isotope ^{18}O. It falls into the shells of marine organisms and further into sedimentary rocks. By analyzing the isotope ^{18}O in the cores of sedimentary rocks, you can make a map of the isotope in time, and hence the temperature. The figure below shows how the ^{18}O isotope content has changed over the past half million years.

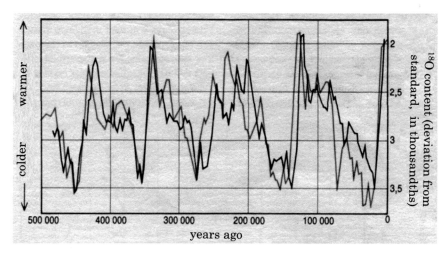

It can be seen that extreme cold Earth periodically, approximately every 100,000 years: in warm periods, the content of ^{18}O in sedimentary rocks was lower and the content was higher in the cold periods. On the right, the vertical axis shows the deviation of the ^{18}O content relative to the standard: the greater the deviation, the higher the ^{18}O content. The two lines, red and black, are almost identical; these are the results of two independent groups of scientists. All of this is a simple commentary on the endless dreary talk of global warming…

Biology

Fascinating Biology

Chemistry is not the whole life.
But all Life is Chemistry...

The most impressive, fascinating and beautiful discoveries are done by biology: it brings what is a cause of delight. Genetic apparatus is delightfully arranged and functions beautifully, molecular machines—enzymes—work flawlessly. And how beautiful ribosome, composed of several molecular machines, each of which performs its function, acts amazingly in concert with others. This enzymatic combine moves on an RNA spiral, accepts the amino acids given to it, select the necessary ones which corresponding to the set code, and synthesizes protein. To get used to something means to lose the feeling of charm, but it is impossible to get used to what ribosome does...

Amazingly arranged and functioning synapses are the structural elements of neurons, where memory occurs, thoughts and ideas are born. The work of a kinesin (molecular motor-cab transporting neurotransmitters from the Golgi apparatus to the synapse) causes admiration. Or the mystery of gene expression... And how the immune system magically works... And, of course, admiration is born by the tremendous consistency of the huge cascade of biochemical reactions that provide the life of the body with building material, energy and performance of various functions. Modern biology is the most effective science, admiring his discoveries in the knowledge of life and its peaks—thinking. And the end of this road of knowledge is not visible.

© Springer Nature Singapore Pte Ltd. 2020
A. L. Buchachenko, *The Beauty and Fascination of Science*,
https://doi.org/10.1007/978-981-15-2592-6_6

Molecular Biological Theater

Molecular biology is a brilliantly staged molecular theater. Genotype and phenotype, ribosome and microsome, nuclease and restrictase, exons and introns, centromere and centrosome, mitosis and meiosis, transcription and translation, chromosome and chromatin gametes and alleles, mitochondria and enhancer, revertaza (reverse transcriptase) and telomerase, kinetochore and mitotic spindle—charming words, and behind them, amazingly beautiful structures—the molecular actors of the theater and the dazzling beauty of the roles of the molcules. There are beautiful words in philosophy, but there they are deprived of sense. In biology, behind the beauty of words and concepts there are functions, the performance of which is life.

Molecular biology stands firmly on the chemistry of a giant reactor, which is any living creature—virus, microbe, plant, animal, human. Its foundation is genetics that is the molecular science, combining biochemistry and medicine, embryology and ecology, paleontology, anthropology and ethnic geography. Being united with the primary Darwinian idea of evolution, this science created a synthetic theory of evolution—the core of life, controlling the arrow of time. This science has revealed the great mystery of life: why it reproduces itself. It has established a part of Earth, where a person and mankind are from; as the person moved and mastered open spaces of the Earth. It has explained why living organisms are so different, why everything that happens to molecules in the human genome echoes in its phenotype and appearance—in the color of its eyes, the shape of its lips, growth, gait, character. What is the source of the disease, whether it is possible to avoid them, how to get rid of them... What is apoptosis and why plants shed their leaves... Why the tiny creature amphibian is 25 times richer than the man in the number of genes, but has no unsurpassed human intellect... Molecular biology proved that man is not descended from monkeys (this persistent slander against Darwinism is born and supported by the Church and its dignitaries), that the genetic branches of humans and apes diverged six million years ago and since then have made an independent movement for various locations...

And there is still much that goes beyond the canonical science... Example: recently mummy of a man who lived in the Stone Age, about 5300 years ago, was found frozen in the high-mountain ice in Switzerland. In addition to the circumstances of his death (what, how and where were injured, what he ate, and what he died of), the structure of his mitochondrial genome was determined. It was found that its genome was different from the genome of modern people and that it belonged to the extinct genetic line of a modern human.

More such genetic evolutionary histories come to us from the paleontological finds of the remains of ancient, almost magical creatures that once inhabited the Earth and long-extinct species like the archaeopteryx and the brontosaurus… The paleontological chronic of life becomes molecular genetics history. The past is no longer a secret. And genetic so opens the door into past… It's magical: using the past to see the future…

Genetics: Man as a Treasure of Relics

"View of the world through the window of the genome"—so the brilliant Russian scientist-geneticist Sverdlov called his book on genetics. This is a window into the inner world of life, the world that is charmingly beautiful, delightfully harmonious, and wisely arranged. It is based on one molecule—deoxyribonucleic acid (DNA). The molecule is simple and chemically even primitive: it is composed of four nucleotides connected in a polymeric polynucleotide chain by phosphate bridges. A nucleotide is a phosphate group joined to a nucleoside. A nucleotide is a nucleic base linked to a sugar molecule—ribose. DNA has four bases: adenine (A), cytosine (C), guanine (G), and thymine (T). The polynucleotide chain as a sequence of nucleotides is denoted by these letters: AGCTGA, etc. The two chains are combined, weaving into the correct double helix. The DNA molecule is usually understood as a magic double helix, the icon of the biology of the twentieth century. The total length of DNA molecules in the cell reaches 1–2 m, but they are folded on histone proteins, reducing the size of millions of times. All this information is the forgotten information from the school textbook. However, education is what remains after all that has been learned is forgotten…

DNA is a smart molecule, self-replicating one… The process of its reproduction is called replication. The smart enzyme—helicase—moves along the DNA molecule and unravels the spiral. It is followed by even more intelligent enzyme—DNA polymerase, which uses the old DNA chains to build new DNA threads, attaching the needful nucleotides step by step. Thus molecular machine "sews" a new thread of DNA and corrects mistakes that polymerase occasionally admits in his "sewing" business. In the correct DNA helix, there are A in one thread against T the other and C against G. And we must also remember that DNA is complexed with proteins, forming chromatin; the latter forms chromosomes—particles that carry all the heredity, biological portrait, and history of the life of each and any creature. A human is known to have 46 chromosomes. The difference between men and women lies only in one of them.

Sections of the DNA chain, i.e., blocks of sequences of letters from the alphabet A, C, T, and G, make up the gene… However, not any section is a gene, but only one that encodes a certain protein. The totality of all genes is a genome. The word "encodes" means that this gene stimulates the synthesis of the selected protein and it is responsible for its appearance. It is here the bridge lies between the genome (or genotype) and phenotype (appearance and its external features).

The genetic code is a set of biological features inherent in the genome, and genetic coding is the operation of protein biosynthesis, i.e., the translation of the genetic code into protein content. Any sequence of three letters (three nucleotides) in the DNA chain is a "codon," it corresponds to one amino acid in the protein to be built a given gene. Coding in terms of common sense is a strange process. It is happening so. There is a clever molecular machine enzyme, the RNA polymerase. The machine moves along the DNA chain as along a matrix and synthesizes a new chain—ribonucleic acid (RNA); it is called matrix RNA, and the synthesis is called transcription, rewriting the DNA code in the RNA code. Recall that each nucleotide in RNA is different from the DNA nucleotide with just one extra oxygen atom in the ribose; therefore, RNA acid is called ribonucleic one, and DNA acid is called deoxyribonucleic one.

Let's come back to the matrix RNA (mRNA). There is another wise molecular machine—a ribosome, which moves along the mRNA chain. It synthesizes a protein chain, attaching corresponding amino acids to each other. Another molecule of RNA called transfer RNA, (tRNA), brings amino acids to the ribosome. Protein synthesis by mRNA is called translation, i.e., translation of DNA code through mRNA into protein code.

Why it is so complicated? After all, a sharp-sighted and wise Creator would arrange this whole hustle much easier: he would immediately organize the synthesis of protein on DNA, bypassing mRNA, tRNA, polymerases, and ribosomes. But the Creator had nothing to do with it. There was blind evolution, which slowly did what he could, amassing all was created, acquired, functional and keeping it. The whole structure of the genome and all the events that come out of it are a million-year-old memory of evolution, its relics. In this sense, man is a treasure of relics…

In fact, the situation is even more cunning. DNA polymerase builds a primary RNA molecule on DNA strands (it is called a primary transcript). It has areas that are able to encode protein (exons, their size is about 1000 base pairs), and areas that are unable to do it (introns, their size—from 50 to 20,000 base pairs). Further, the primary transcript (this is immature RNA, on which protein synthesis does not go) is converted in the cell enzymatically into a real, mature matrix RNA (mRNA), on which the ribosome already makes the protein. The conversion of primary RNA into mRNA is called splicing. In this process, introns are cut out and exons are sewn into an mRNA chain. And it comes in a special particle—spliceosome, composed of complexes of RNA with proteins. And this fact was a major event in genetics: it turned out that RNA is able to work not only as a translator of DNA structure in the protein structure but also as an enzyme. Immediately there were ideas that the world of RNA was the primary in evolution, which later gave way to the modern world of DNA.

In fact, the activity of RNA is even more cunning. There is a type of splicing called alternative, where the primary transcript loses not only introns but also some exons. Then there are different mRNA; some of them combine some exons, others collect the best ones. Hence the diversity of mRNA and the variety of proteins needed to different tissues and organs. Moreover, although introns do not encode proteins, their role is paramount in the regulation of genes, in such phenomena as the shuffling of exons, the appearance of jumping genes and enhancers—regulatory sequences of ribonucleotides. And then there are transposons… But let it be for intrigue…

Music of Genes

The genome is the guardian of heredity; it is stable, autonomous, and reproducible in generations. All mankind—and it is already seven billion individuals—is set of 25,000–30,000 genes. These are genetic notes. And as magical enchanting music is built on seven simple notes, so a huge chemical world is created by 16 chemical notes. And all mankind sounds genetic notes, magical music of genes. In reality, 99% of the genome of all people is identical; all the difference is 1% of the genome. But what a variety, what a diversity… A romantic philosopher says: each of us has something of our own, mysterious. The clever geneticist answers: these are genes, my property, my everything; in them my sympathies and hostility—those that at first sight.

A genome is a wise child of evolution. It combines the incongruous—stability and variability, rigidity and plasticity. The coding part is only 1.5% of the genome; everything else is the totality of non-coding genes, gene-regulatory, gene-managers, and gene-officials. They control the coding genes, they give signals to the coding gene to start its work (this work is called gene expression). They determine in which organs and tissues and at which point expression should be switched on… And which gene should work… they turn on the gene or, on the contrary, switch it to silent mode. Thus, cancer cells can be in silence indefinitely long, but it may happen (and it is not known why) that switch genes will command cancer cells to express their genes. And that is a bad command. In the same way, the command for apoptosis—cell suicide—is given.

The command to the expression of the encoding gene is given by noncoding, regulatory DNA sites with switch genes. They can be 10–50 times larger than the encoding gene itself. Gene managers command through the synthesis of special proteins that bind to special areas of DNA that are called enhancers. Joining the protein to the enhancer triggers gene transcription, i.e., puts it into operation. Every gene has at least one enhancer. Usually, it consists of several hundred nucleotides and can be located on either side of the gene. Even inside it. Some genes have multiple enhancers; each of them acts independently and regulates gene expression in different cells, in different parts of the body and at different times of its life cycle. There is a lot of mystery and therefore magically attractive in all this… And it is very similar to the life of people in bureaucratic society…

Relics and Paradoxes of the Genome

We have already noted that human is a treasure of relics. Moreover, most of the cells in humans are alien; they are inherited through evolution from ancient bacteria and lower prokaryotic organisms. Some enzymes—synthase and polymerase, RNA and DNA are also the inheritance; they all retained structural, chemical, and functional properties only slightly modified by millions of years of evolution. And DNA of mitochondria—organelles, supplying the cell with energy—is a wonderful mixture of prokaryotic and eukaryote genes that are genes of lower and higher organisms. It stores the memory of human evolution (more on that later).

The human genome has three billion pairs of nucleotides and about 25,000 genes. The simplest creature, the mycoplasma *M. genitalium*, has a minimum genome of only 470 genes. However, it cannot provide for himself completely, so it parasitizes in a human. The bacterium of walking pneumonia *Mycoplasma pneumonia* is no more complicated; it has 689 genes and 200 enzymes. And it also parasitizes in human. The bacterium *Helicobacter pylori* live in the stomach of people suffering from peptic ulcer disease; its genome has 1.7 million pairs of nucleotides and 1590 genes. Drosophila is the fruit fly, measuring 3 mm: its genome has 165 million base pairs and 13,000 genes; it is only half the size of a human genome. And what is the distance in phenotype! Where this fly and where people… By the way, drosophila occupies a dear place in genetics; as noted by E. Sverdlov, it is considered the queen of genetics.

Japanese puffer fish is deadly poisonous fish, fish of fearless samurai who eat it without removing the poison. It's like a deadly duel: 0.5 chance to survive. The genome of this fish is 365 million pairs of nucleotides (9 times less than in humans), but the number of genes is the same. That is, its genome is much more compact than the human one; it has fewer introns and regulatory genes. The genetic branch of this fish has separated from the human branch about 450 million years ago. The mouse genome 99% coincides with the human genome; their genetic pathways have diverged recently, about 75 million years ago. Chimpanzees and humans (their genomes are 98% the same) make their own and independent evolution for more than six million years. A shark and a man shared their genetic destinies about 500 million years ago. Somebody believes that sharks do not sick cancer; this means that the gene responsible for carcinogenesis appeared later, after the separation of genetic branches. The dogs were separated later, for they had already inherited cancer.

The genome discovers strange and even paradoxical properties. The full size of the genome is not related to the complexity and level of development of the organism. Amoeba has 200 times more DNA than the human genome. Coding genes are dominated in amoeba, but it almost has no genetic switches; these facts determine the plasticity of the genome. Further, the number of coding genes is not related to the complexity of the organism. So, a human has only 2 times more genes than a tiny worm. However, a worm has only 1000 cells, and a human has 1014 cells, which is 100 billion times more. And these cells are 200 varieties and they produce five million different proteins! And bean cells contain even more DNA than human cells…

All these oddities mean that the evolution of genomes occurred unevenly, non-monotonously, there were periods of rest and explosion—depending on the conditions in which the genome carriers appeared after their genetic destinies had diverged and their evolution went to different addresses. You can build reasonable hypotheses and explanations, but the choice of truth is ambiguous. The truth has disappeared beyond the horizons of hundreds of millions of years gone by. However, geneticists surely go up the trail…

Gene as a Guide-Book: The Bar Code of Personality

Where and when a modern human has appeared, where, how and in what ways he has settled on the earth's continents—these are the questions that ethnogenetics and population biology give answers. The answers were found in the genomes of people of different ethnic groups, races, and nationalities. We have already noted that all of humanity sounds the music of genetic notes. 99.9% of all human genomes are the same, but the remaining 0.1% contains all human diversity. People's differences are measured by DNA markers, genes, or groups common to a given ethnic population. The answers to the questions—where? when? how?—follow from the analysis of DNA markers. The science of this is ethnogenomics.

The best of all markers is DNA placed in mitochondria—tiny containers in which ATP (the main energy carrier in the body) is synthesized. The number of genes in mitochondrial DNA is much less than 20,000. There are 37 unchanging genes and there are areas of the genome that change rapidly at almost constant and known rates. These sites, counting the time of mankind existence, as well as the moments of time when the genetic branches of ethnic groups, races, and peoples diverged, are used as a molecular genetic clock.

The genetics of mitochondrial DNA (it is inherited along with mutations only on the female, maternal line) and DNA of the Y chromosome (it is transmitted along the male line, only from fathers to sons) proved that the common ancestor of modern man was a small population (several thousand individuals) that lived in Africa and on the site of modern Yemen about 200,000 years ago. For the first time, this conclusion was obtained from mitochondrial DNA. So the African population was called conditionally and romantically the foremother of mankind—the mitochondrial Eve—the name taken from the biblical legend about an origin of people.

About 60,000 years ago, a small group of this population (a few hundred, maybe a thousand people) crossed the Bab al-Mandeb strait and went on a great journey to Europe, Asia, and America, a journey lasting tens of thousands of years. Human migration routes can be depicted on the map as clearly as the movement of current people.

The genetic DNA markers are road signs on the highways of migration. Following the movement of the Y chromosome, we will see that M168 marker disappears and the M89 marker appears instead when crossing the Bab El Mandeb strait and moving North through the Arabian Peninsula. Then the road turns to the right and together with the new marker M9 follows to

Mesopotamia and further. Reaching the Hindu Kush, it turns left (marker M45), and when it reached Siberia, makes a sharp turn to the right (M242), crosses the border between the continents and appears in Alaska. Further, it crosses South America with a marker M3.

DNA markers reveal many interesting things from the genetic history of nations. Thus, it was found that Bushmen in southern Africa remain genetically isolated from the rest of the world for 100,000 years. Some part of the Lebanese genome comes from Christian crusaders and the Muslims of the Arabian Peninsula. The ancestors of the aborigines of South America come from people who lived on the site of modern Siberia. Genetic tests of the African trace and its migration are absolutely independently confirmed by other markers—from anthropology, linguistics, and archeology. It is remarkable that they can be traced in the genetic evolution of the bacterium *Helicobacter pylori* accompanying a person (through gastric ulcer). So, in the colonization of the Pacific Islands, there were two waves: one was 37,000 years ago and spread from Asia to Australia. The second, from the same Asian ancestor, departed recently, about 5000 years ago.

And, finally, DNA markers can distinguish one person from another, to establish the relationship of man and his parents (everyone knows the story of the genetic identification of the remains of the Romanov Royal family), to identify people who died in disasters, wars and natural disasters, and to find the offender who left biological traces (blood, hair, etc.; it is now called DNA fingerprinting). DNA markers are a human barcode, his genetic passport. And this is the way to personal medicine…

Epigenetics

Genes know how we live, what we eat, and what we drink… The genome is a fragile delicate creature that reacts subtly to its master's lifestyle. We must realize that the genome is passed on to future generations, and they will remember for a long time those who gave them this genome, just as we remember those whom we received the genetic heritage from. And here is the problem how we will remember—with gratitude or on the contrary—it is put already by many factors; among them, the significant place belongs to a way of life of the genome owner.

The gene is fate; changes in it are twists of fate. And a little predictable one… The more changes (mutations), the steeper turns… However, mutations are not the only mechanism of genome evolution. Chemical reactions of DNA can occur without affecting its mainframe chain. The most significant reaction is methylation, joining of methyl group CH_3 to nucleic bases—adenine, guanine, cytosine, or thymine. Such methylated genome manifests new traits. These traits are inherited and alter function (expression) of genes, as well as mutations. However, unlike mutations, genome methylation and the traits it produces are reversible. This kind of genome evolution has been discovered recently and is dealt with by epigenetics, a special field of genetics that studies the impact of living conditions on the evolution of the genome.

It is said that epigenetics began with the fact that scientists at Duke University (USA) introduced the AGTU gene to mice (with a sequence of nucleotides adenine—guanine—thymine—uracil). The result was a transgenic mouse (called agouti, gene name)—being fat, yellow and sickly, predisposed to cancer and diabetes. All these qualities were inherited. However, when mice producing offspring were put on a special diet—food enriched with molecules with methyl groups—suddenly normal mice were born—gray, healthy, and long-lived. In this case, no nucleotide has changed its place in the DNA chain, although the genome has clearly changed. It's all about epigenetics: with an excess of methyl-containing molecules, enzymes that produce methylation were activated; methylated nucleotides turned off the AGTU gene, it was put to sleep and all its bad qualities disappeared in the next generations.

Cancer and epigenetics—that is the anxious issue. It has long been known oncogenes that trigger the conversion of healthy cells into cancer. In healthy cells they are turned off, i.e., they are in a "sleeping" state. This state is kept by the methylated promoters—a pair CG (cytosine and guanine, united by phosphate group). A number of such pairs in the DNA molecule can be from 200 to several thousands and they are grouped into islands. While they are methylated, the oncogene is silent. But if methylation is suppressed and inhibited (for example, by substances of tobacco smoke or acetaldehyde—the product of the conversion of ethyl alcohol of alcohol consumed), the oncogene wakes up and begins to work actively, starting the cancer process. And this is not speculation; this is a scientific result, supplemented by harsh statistics.

And here is an example from brain genetics. Biochemists at the University of Texas analyzed the genome of chronic alcoholics and found them 163 genes in their brain, and work of these genes deviated from the norm. These genes are responsible for enzymes that synthesize myelin that is a protein covering the axons of neurons in the white matter of the brain and isolates them from other axons. These genes are almost paralyzed by alcohol and do not produce myelin. As a result, nerve impulses of different axons overlap and "close." The targeted delivery of pulses is disrupted, creating chaos in their transmission.

It is the origin of the problems with memory, attention, emotional anomalies, and unpredictable behavior of alcoholics.

Epigenetics is actively developing; it has already been proved that epigenetic changes in the genome (in particular, methylation) are responsible for overweight, for some forms of schizophrenia, and for longevity. Epigenetic changes in the genome depend on the way of life: in this sense, epigenetics opens up ways for each of us to save and preserve the genome in a healthy form.

Telomerase Is an Enzyme of Immortality

Stem Cells: Longevity

Once the author fascinated by telomerase wrote this:

> *The chromosome has an end.*
> *It's called telomere.*
> *There are no genes,*
> *But it isn't a chimera.*
> *It saves the DNA,*
> *Prevents devastation,*
> *Preserve youth of all –*
> *And fly, and cat, and bear.*

Of course, this is not a masterpiece of poetry. And the pathos is greatly exaggerated. First, namely, the telomerase preserves the youth rather than telomeres (it was written for the rhythm only); secondly, it retains youth not all kinds of species, but selected ones only—embryonic stem cells. There are special, short, and repeating, but not coding sequences of nucleotides called telomeres at the ends of chromosomes. The human telomere is made up of hundreds of such TTAGGG repeats (recall that these are the letters of the nucleotide alphabet). DNA polymerase, carrying out DNA replication, does not rebuild these ends. Therefore, each act of replication is accompanied by the loss of ends and shortening of DNA. And then it turns on telomerase, which corrects the negligence of DNA polymerase. It builds the ends of the telomere, restoring the complete DNA chain after each act of its replication.

This enzyme is amazing, original, and unique. First, the telomerase is an RNA polymerase; it is called reverse transcriptase. The biological world is amazing: it is well known that DNA controls the synthesis of RNA (see earlier). So the opposite happens here—the RNA builds the DNA. Hence the name of the transcriptase (reverse code rewriting) has come; hence the idea about the primacy of the RNA world (see earlier) has appeared. Secondly, telomerase is present only in embryonic stem cells—those that have not determined yet who they will be: the heart, skin, bone, liver, brain, or something else… These cells divide quickly and often, they are in need a lot, so DNA replication is performed intensively and there is a huge need for a lengthening of unfinished chromosomes. Namely this work the telomerase does… It is what makes stem cells immortal. Thirdly, fate and functions of somatic cells are already defined, so there is no telomerase. Therefore, somatic cells are able to divide, to double a limited number of times only, about 50 times, then they die. And the DNA polymerase puts a limit, more precisely, its inability to build a telomere. When the limit number of divisions is reached, replication stops, the cell stops dividing and dies. This is the well-known Hayflick law. (By the way, amazing movie to watch as a cell divides, at www.mitocheck.org. It is a living movie and not animation.…)

At one time, telomerase was associated with aging and span of life. But there is no reason for this connection. It is believed that aging is simple to wear and tear of the body. As one biologist noted, "there is no reason for natural selection to put supports to someone whose reproductive age has been left behind." A strong word… But genome is inexhaustible… A family of genes was recently discovered that is in opposition to every stress and aging. These "longevity genes" were called SIR or SIRT (these SIRTs were found in mammals). They encode the synthesis of proteins Sir or Sirt. Thus, Sirt2 affects the processes of cell division through modification of tubulin, which is used to build cell structures. Sirt3 affects the mitochondria, where ATP energy carrier is synthesized. Mutations in the Sirt6 protein gene produce early aging.

It seems that the strategy of Sir proteins is to keep genes silent. So, SIR2 encodes an enzyme that has an unusual property. The DNA molecule in the cell is known to be wound on histones. There are chemical labels—acetyl groups attached to histones that support the desired packing density. If some of the labels are removed, the DNA is wound too tight, so the enzymes that remove its ribosomal DNA are powerless. DNA becomes "silent" in such a dense state.

The work of the enzyme Sir2 is associated with the presence of a small molecule NAD (nicotinamide adenine dinucleotide). And NAD is active in metabolism and is directly related to the nature of nutrition and aging. Hence the direct path to diet: it is found that animals that consume food 30–40% less than the average live significantly longer and healthier, as their metabolism is slowed. This conclusion about SIR2 genes as life expectancy genes has been reliably proven. Next are ideas on how to find drugs that change the activity of enzymes encoded by SIR genes. Then there is a way to fight Alzheimer's disease, diabetes, neurodegenerative disorders, and other troubles. Some of these drugs are already undergoing clinical trials.

Let us return, however, to the stem cells. They are multifunctional (biologists say: pluripotency), i.e., they can be targeted to produce somatic cells of the heart, skin, neuron, etc. Thus it is possible to produce tissues of different organs. And that's what the medical needs: the regeneration and transplantation of tissues and organs. But stem cells are embryonic cells; their source is embryos, human embryos. And here the delicate moral and ethical problems begin. As a result, in most civilized countries, work with stem cells was semi- or even completely banned. Recently, Barack Obama, the President of the United States, partially lifted the ban on those stem cell lines that are derived from excess embryos in clinics that practice artificial insemination. But things change quickly here…

Not so long ago, a major discovery was made—it was possible to transfer the biological evolutionary clock back and return the somatic, unipotent cells again to the start, turning them into stem pluripotent ones. This is a major breakthrough in genetics and molecular medicine with prospects in practical medicine. And this is the beginning of a great road…

Cancer, Carcinogenesis

Evolution did not protect living organisms from the terrible and almost inevitable disease—cancer; moreover, it created it. The probability of cancer throughout life for women is 39% and for men is 45%. The United States loses one person every minute due to cancer. Getting rid of cancer is possible; there are many cases of healing. But the biblical commandment work here: "Saved is saved to the end, doomed is doomed completely…."

There is no unambiguous theory of the origin of cancer; there are several ideas and many experimental results. They can be approximately generalized at two levels—the level of somatic cells and stem cells. Somatic cells are the normal cells functioning at the identified locations: the cells of the heart, liver, blood, skin, etc. the first idea was simple—the source of cancer must be sought in the genes. Indeed, the first human gene carrying cancer damage (oncogenes) was discovered in 1981. Later it was found that the root causes of cancer are mutations in certain genes. They can occur under the influence of toxic substances and radiation due to impaired DNA replication or errors in DNA coding before cell division. Sometimes the mutant gene is inherited.

Regardless of the cause of mutations, they always lead to the fact that cancer cells divide uncontrollably; then they penetrate into other tissues and capture the entire body. Already 350 genes have been exposed as oncogenes, an atlas of cancer genes has been created, which is continuously replenished. Some substances were found that inhibit cancer at the genetic level (gleevec, herceptin). The ultimate goal is to make a genetic map of cancer and find ways to turn off dangerous genes. The goal is worthy, although very remote... But people are dying today...

Another genetic approach is based on chromosomal cancer theory. It was born from simple observation. First, it was found that powerful carcinogens (asbestos dust, naphthalene, arsenic, lead) do not cause mutations. Moreover, it turned out that the doses of carcinogens that trigger tumor formation are a thousand times less than the doses that create mutations. The mutation theory of cancer could not explain why non-mutagenic carcinogens trigger the malignant transformation of cells. That is, there are no mutations, but cancer is...

Secondly, it was discovered that the appearance of cancer is preceded by chromosomal chaos—aneuploidy. Its source is cell division. The correct division provides for the symmetrical doubling of chromosomes and their exact divergence in daughter cells. This property is diplodia. But it is possible that the chromosomes are randomly distributed asymmetrically in defective daughter cells. Cells with an abnormal, incorrect number of chromosomes were called aneuploidy ones. They are not viable and the body destroys them through the immune system. But it happens that one or more of these cells survive and begin to actively divide and degenerate into cancer. There is a pattern: cancer cells are usually aneuploid. According to this theory, cancer-inducing substances should not be called mutagens, but aneuploids. Moreover, it was shown that all carcinogens (including radiation) give rise to aneuploidy.

In fact, chromosomal chaos theory implies that the entire genome is responsible for cancer and that it is unlikely that a local cancer-provoking gene can be isolated. There is a lot of biomedical evidence in favor of this theory; it is inappropriate to discuss them here... But let's note one of them: it is possible to destroy tumor cells by any medicine, but instead of them there are others, which this medicine will not kill. Most likely, both mechanisms—mutations and chromosomal chaos—work together. And that seems to be the reason that you can't get rid of cancer at the genome level. By the way, almost the same situation is with AIDS...

Traditionally, it was believed that the body needs to kill all cancer cells; any remaining cell is able to re-start the multiplication and growth of cancer. Awareness and understanding of the properties of stem cells gave rise to a new strategy: it is necessary to destroy only special cancer cells, which are few. Namely, these cells that divide rapidly, multiply and create what is called cancer. This fact was shown using blood and immune system cells. Those cells come from bone marrow stem cells (there are about 0.01% of these cells). These cells are self-supporting: each of them is divided in half, but then only one of them turns into somatic functional cells. The primary cell remains unchanged and continues to self-sustain through new divisions. In this case, the number of stem cells is preserved, and somatic cells are updated and replenish the cells of those organs that need it. After all, somatic cells are mortal (remember, about 50 divisions are only possible), but stem cells are immortal as far as the telomerase works in them. Stem cells work as factories producing somatic cells, which replace the dying ones. And this is their noble function…

But if the stem cells are converted into malignant (due to random mutations, but every random event has its own causes), then they become the locomotive of cancer and its immortal successors. They will also give birth to somatic cancer cells; the latter go to different addresses, to different organs, spreading cancer everywhere. The new anti-cancer defense strategy is to kill the cancer stem cells. There is the first success: a substance was found and extracted from guayule plants that stimulates the apoptosis (suicide) of cancer stem cells but does not affect normal stem cells. So somatic cancer cells kill themselves because they do not have telomerase!

The general and the most effective line of protection against cancer stem cells seemed to be as follows: it is necessary to find means that quickly turn stem cells into somatic functional ones. It is a method of neutralization of cancer stem into cancer somatic. Well, these ones are not dangerous; they will die themselves because they have telomerase… This way is long but sure…

What Is the Price of the Life?

No, it is not the money… It is an energy price. The Earth receives from the Sun 10^{21} kcal per year; it is believed that only a thousandth of this energy is used in the photosynthesis of 100 billion tons of green mass. It is the primary source of food energy for all life on Earth. Half of all life on Earth and half of the energy consumed on Earth are microorganisms; the other half is given to invertebrates and lower animals. Warm-blooded birds and mammals are only a small part of the living world both on weight-mass and on energy consumption.

An active person needs about 2500 kcal per day, which he receives with food. But every calorie of food takes about ten calories of photosynthesis.

That is, a person costs 10 times more expensive. And if we take into account the energy costs for infrastructure (housing, transport, comfort, luxury, etc.), the energy costs will increase by 10 times, i.e., the energy cost of a person is about 250,000 kilocalories or 300 KW-h per day.

How we spend those 2500 calories that one gains from food? It is clear that the work of all organs—lungs, kidneys, stomach, liver, muscles, brain, etc.—requires energy. But these costs account for only 20% of all energy. The remaining 80% goes to the renewal of the body: the decay and synthesis of cells, biomolecules, enzymes, proteins, nucleic acids. The update of the chemistry of living beings is the most energy-intensive process. Thus, every day the human body renews about 4% of its protein composition, and to maintain the electrical potential of the membranes of neurons spends about 30% of the total energy reserve of the body.

People, as well as any other living creature, are energetically expensive. Moreover, its energy efficiency is woefully low. Why is it so small and why is life so expensive? Why are living things so energy-consuming? And who is the marriage-maker who created such imperfection? The answer is … The energy that comes into the body with food is not used immediately. All of it through a variety of enzymatic reactions is spent on the production of adenosine triphosphate (ATP) that is the main molecular energy carrier in all living organisms. And it is the universal one, it serves all the functions of the body with energy. Nothing is done in the body without ATP. It is formed and consumed continuously; its daily production is half the body's live weight.

There is about 200 g of ATP in the human body at each time. The decay of 1 g of ATP molecules releases about 20 calories, i.e., 200 g of ATP releases 4.0 kcal. To ensure the daily energy requirement of 2500 kcal, it is necessary that ATP should be updated (re-synthesized) 2500/4.0 = 625 times. That is, almost every 2 min the body renews its ATP, its energy reserve. ATP in all living organisms is produced by the same enzyme—ATP synthase (it was discussed earlier). The body of medium (by weight, not by intelligence!) person consists of 10^{14} cells; each cell has about a thousand mitochondria, and each of mitochondria includes 10^4–10^5 molecules of ATP synthase. Thus, the total number of this tirelessly working enzyme is 10^{21}–10^{22}. It works almost the same in all organisms. So this is a consequence of the evolution that created and consolidated such universality and structure, the mechanism of energy supply (through ATP), and the method of synthesis of this energy carrier. The reasons for the high energy consumption in living organisms are that the efficiency of ATP synthase is very low; its efficiency is about 10%. If the living world would be created by the Creator, he would arrange it differently—so that the efficiency would be high, the energy consumption would be low, and life itself would be cheap. But it was not so… There was a blind meaningless (i.e., without any thought) evolution, which inadvertently (foolishly, as one biologist said) created and inherited such an energetically expensive life. Therefore we have to put up with it and live in debt on account of the irrevocable credit of future generations…

Intelligence

Intelligence and Thinking

The mind is a treasure, a mysterious thing, enigmatic, and elusive… Namely, the brilliance of the mind (rather than the shine of lipstick on the lips) creates individuality and dissimilarity of each. The mind is the work of the brain; about it later. The mind has always been the subject of reasoning and evaluation, the subject of admiration and envy, delight and ridicule. Countess Varvara Golovina wrote about Russian Emperor Paul I: "…his head was a maze in which the mind was lost." Indeed, the mind often serves to do stupid and crazy things.

The human mind is abnormal; it stands out sharply from the mind of the rest of the living world. Man and mankind have created a civilization—one that none of the animal communities have created. Bees, ants, and microbial colonies have their own civilization, but it's out of integrals and quantum mechanics. The brilliant superiority of the human mind is the ability to go beyond frames of the obvious, accessible, tangible world, the ability to think abstractly using imaginary images.

Human has two minds. One is algorithmic, it appears, improves and enriches as a result of experience, training, and education. The perfection, depth, and strength of this mind are signs of talent (which is known to hit a target that no one can hit). However, there is another mind that is non-algorithmic, existing independently, uncontrollable. The mind is mysterious and divine; it is a source of sudden insights and guesses. The mind is unexpected, independent, and unpredictable, often beyond even his master. The mind can be high and generous. The mind of a genius hits a target that no one

© Springer Nature Singapore Pte Ltd. 2020
A. L. Buchachenko, *The Beauty and Fascination of Science*,
https://doi.org/10.1007/978-981-15-2592-6_7

sees. Are the clues of the mind on the roads of science or they are beyond science and are not subject to science; it is an intriguing question; about it—a little later...

The algorithmic mind is built through learning, cognition, and education. Education is the acquisition of knowledge and algorithms of their use, and the mind is their receptacle. Knowledge is the food of the mind: children learn the multiplication table, sparrow learns to fly... And so on...

The inhabitants of provincial Africa are brought up on their own, special musical rhythms, the music of Bach and Chopin do not inspire them. However, Africans, educated and trained in Europe, are susceptible to Bach and Chopin, as native Europeans. The inhabitants of Polynesia and Europeans almost do not understand each other in spite of their full genetic identity. Different algorithms, different algorithmic minds, different cultures... Habit is an algorithm that is extremely robust and well established due to repetitions.

The great Sophocles was right in claiming that the mind was the first condition for happiness. Russian Nobel Prize laureate Ivan Pavlov wrote: "the main task of the mind is the correct vision of reality, a clear and accurate knowledge of it." He insisted: "...concentration of the mind is strength, but mobility, running of thought is a weakness." There is no need to talk about the advantages of a strong, developed, educated mind. Aristotle argued that the difference between an educated person and an uneducated person is the same as between a living person and a dead person. This, of course, is a strong exaggeration (like everything on what Aristotle insisted). Closer to the truth was Theodore Roosevelt's remark, one of the American presidents: "an uneducated person can only rob a freight car, and a University graduate can steal an entire railway."

An algorithmic mind is available for improvement and training, filling and expansion. Known joke: "Even if the knowledge is given to you for free, the container should be always bought by you." Everyone wears this container on their shoulders. Immanuel Kant, a controversial philosopher, gave wise advice: "have the courage to use your mind," implying everyone's concern for the wealth of your mind. But the advice is also known: the mind becomes more if it is not spent...

The non-algorithmic mind is much more intriguing; it is inimitable, unique, and non-reproducible. Often it is simply unconscious, it works and creates regardless of its owner, in addition to his will and condition. This is the mind of a genius that goes beyond knowledge and algorithms; its power is beyond the algorithmic mind. Genius cannot be reproduced, he is unique, inimitable. Like a chess game...

The world of thinking, when it goes beyond knowledge, becomes unsteady, becomes a world of mirages. And this is its charm and risk—its innuendo and incompleteness. Confucius warned about this: "it is absolutely useless to study anything and not to think about what you have learned. It is dangerous to think about something without studying the subject of reflection."

Creativity is going beyond the known knowledge. Scientific method should control creativity. It is strict: the invented things and the world of mirages should always be checked for illusory; any guess of mind should be checked for reliability and scientific credibility. The principle should be firmly held: "before stepping on something, make sure it's not a rake." There is a reliable criterion of truth that is the reductionism. It defines the boundary between truth and mirages, between science and pseudoscience. Both the algorithmic mind and its crown—the non-algorithmic mind—are two sources of intellectual wealth of man and mankind, two driving forces of science and civilization.

Brain

Russian writer Anton Chekhov insisted that everything in human should be beautiful... And still the most beautiful is the brain, the seat of the mind and consciousness. And this is the most fascinating, the noblest object of scientific knowledge that is both self-knowledge of the mind and self-awareness of feelings... This is a macro-reactor of enormous complexity and delicacy, in which a huge number of chemical reactions are carried out, producing a synthesis of molecular structures that form memory, emotions, and the entire control system of a living organism. The brain is a reactor of key importance in all the chemistry of living things.

Outwardly, it is not beautiful. Laid out on a plate, it looks like 1.5 kg of gray oatmeal. It is 90% water, but 150 g of dry matter is the whole genius of mankind, all the riches of civilization created by this genius. It is a miracle of thinking and delight of feelings; it has the brilliance of mind and charm of personality, the value and uniqueness of each...

The brain's energy output is about 20 W, it is a dimly glowing light bulb. However, this "light bulb" consumes about 20% of the body's energy. However, the baby's brain consumes, even more, about 50%. More than 80% of all genes work on the brain and its functioning, i.e., genetic evolution took care of the consciousness of living beings in the first place. How the brain works, how it works—this is aimed neurophysiology, neurogenetics, and neurosurgery. Modern brain science has created excellent technologies and smart tools to see how the brain works, for example, positron emission spectroscopy, nuclear magnetic tomography, and clinical neurosurgery.

The surface of the brain is a complex and seemingly disorderly landscape of hills and valleys. The cerebral cortex is the gelatinous blanket with a thickness of 2–4 mm and filled with neurons. It is a gray matter; it defines perception, thinking, emotions, and actions. (Joke about one minister tells that the brain he got from an oak.) By the way, anthropologists have found that the last 30,000–40,000 years the average volume of the human brain decreases. What does that mean? One of the reasons is people use it less and less… What is not used will die… It is the law of evolution.

In animals with a small brain, its surface is smooth. If you smooth the human cortex, its area will be three times larger than the inner surface of the skull, i.e., the only way to put the brain in a small cranial space is to put the bark folds. Folds are bizarre and not random: the pattern of folds is almost constant for all people. It is formed during development and retains its shape throughout life thanks to a network of nerve fibers—neurons that germinate and hold the brain mechanically similar to elastic threads.

The cortex consists of rows of cells stacked in a multilayer (5–6 layers) cake. Layers may differ in thickness and composition. For example, in the cortex areas occupied by primary information, the thickest layer is the fourth, and in areas responsible for the movement is the fifth. In areas where memory is concentrated and thinking occurs, the third layer is the most developed. The distribution of cells in layers divides the cortex into specialized areas, which are the map of the cerebral cortex.

There is an idea and, moreover, there is direct evidence that talent indicators are hidden in the cortex map. A clear connection is found in people who are constantly exercising in something. Examples are professional musicians; their brains have highly developed the fifth layer of the cortex, and this is different from the brain of non-musicians. Brain activities of chess players differ: amateurs and beginners have active temporal part of the brain (thinking) during the game, but professionals have active frontal and parietal cortex (their long-term memory is stored and extracted from).

Today you can observe the human brain during life (magnetic resonance imaging) and reproduce its three-dimensional structure with the help of a computer. One of the findings is the detection of significant differences between the cortex form of healthy people and people with mental disorders. The brain of people suffering from schizophrenia has less pronounced folds and this may mean that the source of schizophrenia is not only the neuro-chemistry of the brain but also the conductivity of neurons. People with autism also have abnormalities in the form of crustal folds: some of the furrows are deeper and shifted relative to their position in a healthy brain. Scientists are inclined to believe that autism occurs as a result of the improper formation of connections in the cerebral cortex. Advances in mapping gave rise to an interesting idea: the map of the brain can distinguish the cortex of genius from the cortex of the offender. The idea is naive: after all, the brain is a tool for thinking and generating any thoughts—good or evil, noble or criminal…

As it was already mentioned, the human genome mostly (about 80%) works for the brain. Thus, the gene HAR1 encodes the synthesis of small RNA, which has 118 differences between humans and chimpanzees. HAR1 was found to affect the development of neurons in the cerebral cortex between 7 and 19 weeks of intrauterine life of the future person. More than a dozen of such genes that critically affect neurons have already been found. This is just one example of what modern brain genetics can and does. It had invented methods of transgenic knockout technologies. Their idea is to use the biochemical methods to add, transfer new genes, or turn off someone, inactivate the specified genes of the genome, and then analyze the consequences of these operations. In this way, to determine what the gene is responsible for, what it does, how it controls the body. These technologies have opened up a wide road to understanding the molecular functioning of the brain.

Is it possible to see a working neuron in action? Yes, an opto-genetics is capable to see those combining genetic engineering methods with optics. After all, the language of neurons is electrical signaling with a voltage from +40 to −70 mV. These signals and their movement through the neutron can be made visible using fluorescent dyes. These dyes change the color or intensity of the glow depending on the voltage of the electrical signal. And they can glow or, conversely, extinguish those ions (e.g., calcium), which create a signal due to the difference in concentrations. So it is possible to see the neuron and the electrical signal running through it.

But you can achieve more and observe the given neurons. For example, it is possible to see dopaminergic neurons—those in which dopamine is synthesized. Dopamine is a neurotransmitter, the lack of which gives rise to such disgusting phenomena as Parkinsonism and Alzheimer's disease. To do this, the gene switch that activates the genes of dopamine synthesis, at the same time should include a gene encoding the dye-protein. And this gene can be implemented by genetic engineering. As the result, only the dopamine-mediated neurons would be painted. And then you could see how they work.

And it was done… The gene of dye was taken from jellyfish, which produce a green fluorescent protein (this protein is widely known to biologists and chemists). This method was tested on fruit flies, and images of certain neuron groups in the working brain were obtained. In the same way, it was possible to see neurons carrying information about the smell.

It is clear that the opposite situation can be done, for example, to activate neurons with light. In studies on mice, optical fibers were used to transmit light directly to the neutrons producing hypocretin. It is a neurotransmitter, a small peptide molecule related to sleep, i.e., to the rest of neurons. Indeed, light stimulation of hypocretinergic (hypocretin-producing) neurons awakens sleeping animals.

The Language of the Brain Is a Coherent One

The gene that determines the growth of the body and hence brain volume was recently found. It is responsible for the synthesis of a protein called pericentrin, which stimulates cell division and increases the body mass and brain volume. The volume of the brain was known to be directly related to intelligence indicator—a popular IQ coefficient.

The brain of adult holds about ten billion (10^{13}) neurons, and each has about 1000 contacts (synapses) with other neurons. Known joke: electrical engineering is the science of contacts. It is no exaggeration to say that the science of the brain is the science of contacts, the science of synapses. The synapse is the crown of evolution, the most fascinating work of art.

The language of the brain is the language of electrical signals. Thousands of them every second enter the brain by neurons and take these signals synapses. They transform them into sensations and feelings, into thoughts and ideas, into memories and actions. A web of neurons combines synapses into a huge molecular system of memory and generating elements. It is in the synapses laid the code of memory, thinking, and consciousness, i.e., molecular chemical mechanisms of reversible conversion of electrical signals into perception, knowledge, and behavior (see below).

There is a wonderful way to eavesdrop on the language of the brain and even talk to him. This way is microelectrodes, which are introduced into neurons. Thus, in experiments with mice, 260 microelectrodes were introduced simultaneously and electrical activity (signals) from 260 neurons of the CA1 field of the hippocampus (the region where memory is formed) was recorded. Next any effect (frightened sound, a sudden flash of light, etc.) acted on freely behaving mouse and the testimonies of microelectrodes were recorded, i.e., the signals of neurons. All neurons responded and their responses were not chaotic. The signals were found to be generated by packets that were subgroups of neurons; they were called neural cliques. These were neurons that collectively respond to a given event. That was, the neurons worked in unison, synchronous, orderly, and also simultaneously gave the answer synapses. Communication with the mouse was not possible, but the person on the operating Neurosurgery was able to communicate at the table. He could be asked questions and received commands, and neurons of his brain, which introduced microelectrodes, responded to electrical signals. And again it was found that these signals were consistent in groups of neurons and that they were coherent. This technology is a way to establish the nature of consciousness, to establish a correspondence between consciousness and behavior, on the one hand, and groups of neurons and the electrical signals generated by them, on the other. This is a difficult and delicate way, but fascinating and therefore inevitable. There is also a practical benefit: this technology helps neurosurgeons to find the sources of motor pathologies (for example, in the surgical treatment of torticollis).

The brain is a vivid example of a coherent organized and self-organizing chemical system in time and space. The chemical activity of enzymes and, as a consequence, the electrical potentials in the system of synaptic membranes and neurons are perfectly synchronized and coherent. The extent of coherence (i.e., the size of the synchronized brain regions) varies at different levels of brain functioning.

The normal state of the brain is ordered. Perfect order, perfect coherence is the generation of new ideas, thoughts, the ability to reflect and create, and it is characteristic of a talented and brilliant mind. The higher the coherence is, the brighter the creative potential of the mind and its genius are. As a random set of colors is not a painting, so a simple set of neurons and synapses is not a guarantee of the mind. Intelligence is a high degree of organization of neurons and synapses into coherent ensembles. We add that it is not only in genes, it is also in hard training, in constant training and education of the mind...

The brain and consciousness are subordinate beings: the brain governs consciousness, the consciousness governs the brain. When control is turned off, random, wandering in the brain electrical signals neurons visit randomly different synapses and cause in the mind stored there chaotically related (more precisely, incoherent) images and visions. And this is the source of fantastic and often senseless dreams of the sleeping brain. In fact, "the dream of the mind gives birth to monsters..." However, the Russian poet Mikhail Lermontov wrote another: "These dreams are a strange thing. They are a double of life, and often better than real life." Both above quotations are true...

Molecules of Mind and Genius: Relay of Charms

In the story of Russian writer Arkady Averchenko, some person asks: "What do you eat that you are so clever?" This is a crafty question: the mind does not depend on what we eat; however, the mind molecules do exist. Memory, remembering, and thinking are chemical processes; these are processes of transformation of molecules in synapses that are chemical micro-reactors of mind. Synapse is the highest element of evolution, and the system of quadrillion controlled synapses is its highest achievement and its crown.

All signals that the brain receives—from all sources, including the brain itself—are converted into electrical potentials. The transfer of potentials occurs through the neuron channel, in which the concentration of sodium, potassium, or calcium ions changes and this change (gradient) is transferred as a concentration wave. The electrical potential (a pulse of 20–70 mV) during milliseconds passes the way from the start, where it was born (for example, in the retina from light or in the skin layer from the burn), and reaches the synapse. The main events that determine thinking occur here. They are simple, but the difficulty lies in the fact that they are a great many. They are diverse, each of them has its own indispensable function and they are composed of some processes: remembering or generating feelings and thoughts. In the science of a brain, the synapse occupies a central place; it is the support of this science. In 2008, scientific journals published 49,000 articles on synapses and their functions. Each of them had new information; each of them was a new step, big or small, on the way of understanding the secrets of the brain.

So, the sig inhanal came to the synapse... It changed the ion concentration in the presynaptic region. There are already prepared microdrops (vesicles) of special molecules—neurotransmitters; they transmit the relay of signals through synapses to other neurons. There are two types of neurotransmitters. The first type includes those that were synthesized from amino acids (each neurotransmitter—from its acid). Their names (acetylcholine, norepinephrine, glutamic acid, serotonin, γ-aminobutyric acid, and dopamine) are known to those who suffer from neuropathologies (for example, a deficiency of dopamine and disorders in its synthesis is a source of Parkinsonism). Those who do not suffer and do not know these names may live inhappy by ignorance... Neurotransmitters are synthesized by enzymes in the cytoplasm; they are packed into vesicles (a thousand molecules in each) and live in the presynaptic region. (In passing: caffeine consumed with coffee also works as a neurotransmitter. It stimulates the speed of decisions—it increases by 1.5 times, but the error of decisions also increases—almost 2 times. Choose a dose...).

The second type of neurotransmitters is cyclic low molecular weight proteins (called neuropeptides). They are also synthesized by enzymes in the body of the neuron (in the endoplasmic reticulum) from different amino acids. Next, they are also packed in vesicles; packaging occurs away from the synapse, in the Golgi apparatus. Further, these vesicles are loaded on the transport protein kinesin, which drags them on his transport—the filaments (proteins) and delivers in the presynaptic space. And all this, of course, are traces of evolution: it is very unreasonable to synthesize something in one place and to use it in another remote place. The Creator would not do this: He would provide both synthesis and use in one place, getting rid of the worries of packaging and transportation.

After arriving the signal changes the ion concentration (e.g., calcium) in the presynaptic region. Change of charges destroys the vesicles selectively depending on the magnitude of the signal and the ion charge and releases the selected desired neurotransmitters. They interact with the protein molecules inserted in the synaptic membrane; these proteins are receptors that receive signals. As a result of interactions (chemical reactions!!!), the shape of molecule receptors is changed: their conformation, rotations of atomic groups, the angles between chemical bonds, the lengths of bonds, and the charges on the atoms… Atypical (but not unique) mechanism of chemical intake is protein phosphorylation: neurotransmitters switch on enzymes (protein kinases) that produce a reaction of protein receptors with the adenosine triphosphate (ATP). In this reaction, the end phosphate group of ATP is transferred to the protein hydroxyl group OH; this phosphorylation reaction changes the conformation of the protein receptor. The transformation of receptor molecules has two functions. First, it opens the ion channels in the synaptic membranes, through which the electrical signal is transmitted further to another neuron. And not just transmitted, but also amplified (there are such mechanisms of amplification in the synapse). This relay of signals lasts about a millisecond when channels are open. Then the neurotransmitter that stimulated phosphorylation of receptor proteins is destroyed (again in enzymatic reactions) and the channels are closed.

Secondly, changes in the conformations of receptor proteins in synapses are a chemical element of memory. This modified conformation is a reaction to a certain signal (behind which there is an event or thought). These signals can be repeated (or not repeated); then the conformation can be stored (or not stored) as stable. And this is the key to the stability of memory: will it be short and fleeting, or eternal (the degree of eternity, of course, is conditional…).

Here is an example: in the hippocampal nerve circuits (brain area), signal transfer stimulates the release of calcium ions that release neurotransmitter—glutamic acid—and activates the calpain protein. This enzyme reacts with another protein that is a fodrin lining the inner surface of the neuron cell membrane. The fodrin collapses and exposes protein receptors, making them available for neurotransmitter and subsequent phosphorylation.

And now we will come back to the question of Arkady Averchenko… Yes, there are mind molecules, but they do not come with food, they are synthesized by enzymes, by the body itself, by the master of the brain and mind. The primitive advice of the philosopher Immanuel Kant "you should have the courage to use your own mind" acquires molecular chemical meaning. By the way, brain power is used only by 3–5%, the majority (more than 90%) of synapses is unused… Are they extra? Hardly…

All of the above is a very approximate and simplified scenario of molecular functioning of the brain. Neurochemistry and neurophysiology are huge scientific fields, which successfully employ thousands of talented researchers. The great things they do cannot be described and made available to all in any book. As the Russian poet Ivan Krylov noted: "…we do not write histories…" These are only some of the musical melodies that are happening silently in the brain…

About a Genius

The genome controls molecular and chemical processes of synthesis of neurotransmitters, receptor proteins, transporter proteins, and all other molecules in the living body. Genome determines which genes are switched on and work and which are silent… Namely, genes give commands to the synthesis of enzymes that produce molecules of the mind.

Such genes have already been found (c-fos, c-jun, Krox-20, mKr2, Arg3.1); their list is growing. They give commands (via RNA and ribosomes) to synthesize nuclear proteins (localized in the nuclei of cells) that bind to DNA and activate transcription (rewriting DNA to RNA) of other genes. Those, in turn, give commands to the synthesis of protein kinases—enzymes that provide phosphorylation of protein receptors. There are genes that control the synthesis of proteins in synapses, structural changes in neurons, and the synthesis of protein shells of neurons (myelin proteins).

A phrase is clothes of thought. Language as a way of expressing thoughts is also predefined in the genes. Of course, a thought is nonverbal; it precedes language; although language often works in the absence of any thoughts (this property is typical to a special category of people). It is established that the genes of language (about 70) were found to sit on a plot of seventh chromosome. One of them was identified; it is KIAAO319, it has a direct relation to the dysfunctions of language. Some genes were found on chromosomes 1p36, 6p22, 15q2… This was a very difficult search, not always unambiguous, but noble; its purpose is to help people…

Genes and their ensembles that control the structure and function of the brain are plastic and changeable. It is observed that they react to the novelty factor; this reaction is called memory reconsolidation. The brain fed by new information and the brain stimulated by fresh ideas and strained reflections does not age… More precisely, aging of the working brain is much less than that of the lazy and inactive brain.

So, the origin of genius, i.e., the emergence of a non-algorithmic mind, has a genetic nature. But the appearance of a set of genes that determine the genius of the mind is a random event. And not inherited straightforwardly and accurately. It is the plasticity of mind, genetic accident, and stochasticity of gene combinations that makes the appearance of genius both unpredictable and inevitable. As noted by the brilliant scientist-geneticist Evgeny Sverdlov: "the gene pool cannot be spoiled: geniuses will appear with stubborn and fatally insurmountable inevitability." He also author of optimistic phrases addressed to each person: "You have no talent? Your ancestors didn't have them either? Do not despair, the great combinator and player—Nature can reward your descendants with them…" And it's not just beautiful words; science is behind them. Someone, however, remarked that genius is a disease. It is a pity that it will never become an epidemic…

In each generation of living beings, the genetic composition is shuffled like a card deck. The shuffling mechanism is recombination; it is a fundamental mechanism for ensuring biological diversity including a variety of geniuses. The PRDM9 gene that controls recombination and biological diversity has already been found…

And the last—about the charm… Even an approximate understanding of the molecular and chemical nature of memory and thinking removes the halo of mystique and mystery of the brain. The charm fades and loses colors… But it is replaced by a new one: the desire to understand how the ensembles of chemical reactions are organized, how they are controlled by genes, when and what turns them on, how the alarm occurs… This is a great way of knowing molecular processes, full of beauty and charm… The relay of charms…

Consciousness

This concept is even broader than thinking, almost limitless. The function of consciousness is to understand, to turn desires into actions… Consciousness is the capacity for purposeful actions… The essence of consciousness is in intuition, not in logic… There are dozens of other definitions from psychologists.

In consciousness, all feelings are recorded: love and friendly sympathy, hatred, aversion, passions and fears, anxiety and delight, joy and despair, irritation, resentment and envy, self-preservation and self-sacrifice. And there are many other feelings born from each communication with them and with the outside world. "We humans are children of gods and slaves of circumstances" (Benjamin Disraeli, first Earl of Beaconsfield). Consciousness is often called the soul, or spirit. And this is the territory where philosophers dance. But they do nothing else here. Who really works here are neurophysiologists and psychologists; the first use physical methods to determine the state of the brain, the second systematize the norms of behavior that dictates consciousness. And those and others use the scheme "question-answer," i.e., establish the relationship between exposure and response.

"Intuition is a sense of truth; it leads to the highest realm of the spirit, i.e. religion" (Saint Russian Archbishop Luke Voyno-Yasenetsky). Of course, this is the naive opinion of believing a religious person. In fact, intuition is a synthesis of unreliable and doubtful knowledge, the totality of which gives rise to new knowledge that could be uncertain and unreliable. Sometimes it is checked and often turns out to be true. Such cases are remembered, but other cases, when intuition is not confirmed, are forgotten as unnecessary.

It is believed that the science of consciousness began with Sigmund Freud. Back in 1895, he presented the project "Scientific psychology" (he was 38 years old): "the goal is to turn psychology into real natural science, i.e., to present mental processes as quantitatively defined states of special material particles." Freud fought over the mysteries of the brain and consciousness, but at that time he could not solve them and brought involuntarily all the psychology to the analysis of consciousness in the spirit of low-art and almost tabloid literature. After all, he analyzed results and consequences; the reasons were not available.

But the idea may be admirable. The project caused the author's sharp accusations of self-confidence, megalomania, the intention to become close to God. The great Russian physiologist Ivan Pavlov (Nobel Prize winner) said later in 1906: "The entire human soul can be studied through the method of objective research"; by the method, he understood his technology of studying conditioned reflexes. His thesis was that consciousness is formed in the cerebral cortex, where conditioned reflexes are brought up and remembered. He thought that one can understand consciousness as the system of reflexes. However, the great Russian neurologist Vladimir Bekhterev convincingly and completely refuted Pavlov's ideas: the frog with the removed cortex behaves quite consciously; the earthworm has no cortex at all, but its behavior and search of the suitable soil are quite reasonable. Instincts don't fit into conditioned reflexes. Pavlov's idea to display a map of the brain to the map of the soul failed, despite his powerful mind and his great enthusiasm and perseverance. American historian of science David Zhuravsky called him "an exhibit of the Museum of dead-end ideas in science," and it was absolutely unfair ... In science, 99% of the ideas are dead end, but without them, there would not be 1% of others, bright and successful...

Consciousness is a strange thing; it sometimes turns on, sometimes turns off. Sometimes it does not obey the will (thinking) and rampant independently, and sometimes timidly surrenders... During the Second World War, when Japan was at war with the United States of America, in Japanese aviation there was a suicide squad, throwing their controlled aircraft on American ships. To serve in the squad was the highest honor of the samurai. But those who had families were not accepted. In order not to deprive her husband of the high honor to become a suicide bomber, his wife drowned herself and their two beloved young children. This was not an unconscious act ... Such conscious altruism is possible not only in human society; some animals (bees, birds, ants, and many others) behave similarly in vital and dangerous moments of their existence.

What turns consciousness on or off? You can look for wise and sophisticated answers, but the truth is that there is no simple and unambiguous answer. But it will be... The same David Zhuravsky called his article about Ivan Pavlov: "The impossible project of Ivan Pavlov." And this was an erroneous conclusion... It would be more correct to say: the project has not been implemented yet. But it could be feasible...

Modern neurophysiology and psychology are developing at a tremendous pace. They use the techniques of positron emission, microelectrode technology and nuclear magnetic imaging of neurons, the areas of the brain responsible for behavioral functions of the organism, for conscious action, etc. There is no doubt that consciousness, like thinking, has a molecular and chemical origin. All emotions go back to the functions of protein receptors, neurotransmitters, and signal molecules. The successes here are amazing ones, they are published in thousands of articles and in dozens of books... Example: the activation of a protein receptor (called D2) to dopamine stimulates joy. The gene on which D2 is synthesized (of course, through RNA and ribosomes) was found; it is the gene of joy. The gene of desperation was found. The gene MAO-A was found, this gene makes some people aggressive. It stimulates the synthesis of the enzyme monoamine oxidase, which decomposes neurotransmitters: serotonin, dopamine, and adrenaline. A variant of the MAO-A-L gene was found in aggressive and touchy people; it was called the "warrior gene." Thus a lot of such discoveries have been made. However, more discoveries will be found on the road of knowledge.

Once Erwin Schrödinger, one of the founders of quantum mechanics, said: "Consciousness cannot be described in terms of physics" (meaning, of course, the atomic-molecular level). Then, almost a hundred years ago, it seemed indisputable. Then the object (consciousness) was superior to the cognizing subject. But time changes their places: now the object lends itself to knowledge.

Consciousness and thinking are in unity and in a union. And they are in a common area on the same synapses, in the same neurons, and in the same brain matter. However, there's a limit to thinking. As Stephen Weinberg noted, you can teach a dog to do a lot of smart things, but it will never be able to learn quantum mechanics and learn how to solve the Schrödinger equation. "The virtues of the mind, not supported by the virtues of the soul, are pernicious for man"—it was said exactly like everything that came from the great Einstein.

Artificial Intelligence

This is a favorite toy of intellectuals: philosophers, information technologists, programmers, and mathematicians; it is the subject of abstract mind games. This is the "new mind" that is a contrast to the old mind created by evolution and to the mind that everyone carries in himself. Modern computer science and discrete mathematics are steadily moving towards the creation of a "new mind." The current supercomputer is really a mind that surpasses the human mind. But it surpasses only in problems where clear and strict algorithms are set. This mind is extremely algorithmic, unable to go beyond the specified. And in this sense, artificial intelligence, an artificial brain cannot surpass the mind of the human who created it, who set its algorithms. The "new mind" will not be able to reach the intellect of its Creator. Penrose has written convincingly and perfectly about this problem in his book "The new mind of the king." Recall the finale of his book: a modern powerful supercomputer, embodying all the intelligence of its creators and presented as unsurpassed artificial intelligence, collapsed, unable to find an answer to a simple children's question: "How do you feel?" After all, this question was beyond his algorithms. But the computer is successful in chess with their rigid algorithms.

Artificial intelligence can be taught a lot (for example, to solve a huge system of differential-integral equations), but creativity is not available to him. In this sense, the science of artificial intelligence is the virtual science, in which the goal is unattainable. However on the way to this goal you can get a lot of knowledge useful for other purposes.

Fantastic Brain

Humanity knows examples of fantastic non-algorithmic, creative minds (Newton, Planck, Einstein, and Feynman) and algorithmic minds, working as computers. For some reason, people are more amazed and intrigued by computer minds. They are truly admirable... Here are examples. A professor at the University of Edinburgh Alexander Aitken divided 4 by 47 in his mind as follows: after 4 s he gave the result by digit every ¾ of a second. After 26 digits and a minute pause, he gave out 6 digits; after 5 s of silence, he quickly called another 18 digits and said that the next period of the fraction begins, repeating the previous one. Paul Erdös, a Hungarian mathematician, was already multiplying three-digit numbers in his mind at the age of three. The worker John Buxton was a wonderful computer. It was told that the request to calculate the cost horseshoes, horse nails 140, provided that the price for the first nail 1 of farting and then it for each subsequent nail is doubled, he gave the number 725958096074907868531656993638851106 pounds, 2 shillings and 8 pence. He was asked to square that number, and he issued a 78-digit number 10 weeks later, during which time he continued his daily work, talked on various topics, and dealt with routine matters.

"Blind Tom," a child-slave, born about 1850, almost blind, was unable to talk, but at the age of four, he could play on the piano any songs heard once. At the age of 11 he was tested. He played two completely new compositions of 13 and 20 pages, and he played them exactly. These stories can be continued...

The phenomenal brain possessed Kim Peak, an American born November 11, 1951 (not so long ago, this unique man passed away and the whole world regretted it). He read a page of any book in 8–10 s and memorized it forever. For an hour he read a large book and could literally quote any place of the book years later. Already a year and a half he memorized books that he read word for word. He knew by heart 9000 works. His memory contained all the information concerning 15 topics of interest to him, including world and American history, sports, cinema, geography, space exploration, the Bible, the history of the Church, literature, and classical music. He knew all the long-distance phone codes and US postal codes, the names of all the television stations in the country. In his head maps of all cities of America were enclosed, and he could give a recommendation on how to drive on any of them. He was familiar with hundreds of musical works, he could tell where and when each of them had been written and performed for the first time, called the name of the composer and the details of his life. All, that became him known, was written on disk of his memory forever. He had a mysterious ability to calculate the calendar for any year. When someone told him that he was born on March 31, 1956, Kim a second later said that it was Saturday at Easter week.

Here the brain works like a computer, according to a given algorithm, and the elements of the computer are synapses. And this ability of the brain is almost unrelated to intelligence, to the creative potential of the brain. Here, thoughts, i.e., electrical impulses between neurons, run along ready-made and previously laid paths, while the creative brain paves new paths.

History

Charm of Myths

History is the bloody path of misery and suffering of humanity, a gloriously beautiful path of civilization and progress, the mysterious trajectory in the unknown and unpredictable future… It is the bell of lives and events, a documentary history of the past, arranged in chronological order. Like an encyclopedia that is a mind arranged alphabetically. It attracts, excites, it is worthy of empathy. But you cannot escape from the thought, which was formulated by someone from smart people (not without humor): historical science is fed by three sources: myths of the past, gossip of the present, and fairy tales about the future. What about the historical documents? But it is different, it is not a science, it is a chronicle. Chronicles are not always true stories. Russian historian Nikolay Karamzin admitted that "under the name of Chronicles, historians themselves were deceived." Even in memoirs, the criterion of truth is elusive, and narcissism is almost inevitable. Every book has an author, but there is an author always present in the book. History has only one thing in common with science: neither one nor the other has any notions of morality. But for various reasons: history ignores them, and in science, they simply do not exist…

© Springer Nature Singapore Pte Ltd. 2020
A. L. Buchachenko, *The Beauty and Fascination of Science*,
https://doi.org/10.1007/978-981-15-2592-6_8

The History in Persons and Motivations

Let's remember a recognized authority—a famous Russian philosopher Nikolay Berdyaev and his book "Sense of history." He wrote in this book about the Italian Renaissance: "the essence and greatness of the Renaissance is that it failed and could not succeed." "The basis of history lies in the form of religious consciousness." "History could not solve the problems of the individual fate of man… and so the history must end." These strange and ridiculous statements can be found on every page of his book… These phrases are high-flown nonsense, only a person enchanted by a web of luxury words could declare those a brilliant revelation.

Let's consider a posthumous history of Pyotr Kakhovsky (1799–1826), one of the five executed Decembrists. During the uprising of the Decembrists (December 14, 1825), he killed General Mikhail Miloradovich and wounded two officers on the Senate square. Myth-makers historians have made him a national hero, a fighter for the people and for the ideals of freedom. But the motives that led him to the Senate square were very different, those were personal grudges only. He was demoted from the military service and was stripped of the cornet title "due to obscene and violent behavior, non-payment of debts in a candy store, and laziness in the service." Only his revenge was proved to play on the Senate square. The payback was unjustly cruel, but it is a story, not a myth. And maybe it is a myth. In history, there is nothing unambiguous, except the facts of the incident, and search of the historical truth never was successful… Estimates of historical persons are ambiguous, ambivalent, and contradictory also. History is full of events and facts turned into a myth.

But there are opposite opinions. Russian writer Andrei Platonov: "history is salvation from oblivion." Nobel Prize laureate Alexander Solzhenitsyn: "I wanted to be a memory. To be the memory of the people who suffered a great misfortune." This is the real history that is a chronicle of events, troubles, tribulations, and tragedies of the people of the era, built by the destinies of millions of people. Their faces were sad and happy, broken and shot, honest and mean-spirited, crystal clear, and disgusting. Russian writer Sergei Dovlatov wrote: "History is a fairy tale, slightly decorated with the truth." It was sad humor, because history is a deadly spectacle in which all mankind plays, and all people—both as actors and as spectators change their places. There are heroes and anti-heroes, their places are also changed. There are persons that are a hero for some nations, but a disgusting monster for others, a genius for some nations and a miserable nothingness for others. There are idols and graven images…

And everywhere the passion fatal,
And from the fates, there is no defense…
(A. S. Pushkin)

Motives and driving forces of history… History is driven by two things: human needs (both personal and corporate) and human thirst (this is also a need, but excessive). There are many people who want to be kings, presidents, and presidents of anything. But there are few vacancies, the competition is huge, the struggle is fierce… Everyone wants to have his place and play his role (modest or loud—to the extent of his ambition) in historical events and performances, large or small… There is everything in history: vanity and greed, meanness and altruism, stupidity and cowardice, madness and prudence… Desires are the most humiliating of addictions, and the only struggle in which it is pleasant to lose is the struggle with temptations. Means to satisfy temptations are capture, appropriation, deception, and violence. Violence is the lever of history.

"Boiling water of history sweeps away all obstacles and dams" (Igor Guberman). Usually aggressive and extremely ambitious people (Alexander the Great, Vladimir Lenin, Napoleon, etc.) raise a storm in history rather than in a glass of water. Geneticists have discovered the Mao-A-L gene, which is responsible for the synthesis of the enzyme mono-amino oxidase. But this enzyme does the dirty work carefully: it destroys serotonin, dopamine, and other neurotransmitters that regulate psychology. It turns out that ambitious aggressors are not guilty, their genetics are to blame: they are just mentally ill people. And humanity obeys their wild desires as spellbound instead of treating them. In passing, we note: the ideas of Communism are absurd for they are contrary to the laws of genetics that is rigorous and accurate science.

In the history of any country, there are so many abominations and blood, intrigues and provocations, lies and treachery, meanness and cruelty, abuse and hypocrisy, treachery and slander, betrayal and flattery, murder and bribery that it gives rise to disgust; it is monotonous and boring unethical. "The story drags its triumphal chariot over the millions of corpses" (Karl Marx). Russian writer and historian Nikolay Karamzin wrote that in the history "lust for power awakes to have temptations and machinations, and directing greedy eyes on the inducement of the rebellion and the bloody massacre." History should be not studied, but investigated…

There are only interests in history. Yesterday's allies become enemies today; enemies become allies. As noted by Mark Aldanov, the court of history is a concept for the morons, it does not exist. History is not a judge, but the court of historians is changeable, their judgments are false. What was true yesterday turns out to be a lie today?

Is History a Science?

This question is hardly offensive for historians that are tireless and honest seekers of truth in papers and relics of the past. It was put by historians themselves. It was a title of Sergei Mironenko's report at the Russian-French meeting of historians in Moscow 2010. There are amazing contradictions. On the one hand, historians are fairly intolerant when the authorities force them to write a "good" story and often resort to repression when historians reveal a "bad" story when thieves' raids of gangs of criminals on foreign lands are declared as national state victories. Compare the history of the Napoleonic wars, written by French and Russian historians; there is everywhere half-truth, i.e., the worst lie. And each of the stories presents itself more honest and nobler than the other… On the other hand, the participants of the Moscow meeting in the final document officially condemned those intellectuals who "create unscientific constructions, interpret the events of the past not on the basis of scientific approaches." The urge to impose concepts and interpretations is indestructible as far as a desire to made myths seemed to be delightful. Nikolay Karamzin delicately called it "historical free-thinking."

Soviet historians have taught until recently that communism is the bright future of all mankind. Now they teach that it is cancer from which mankind was cured at the cost of terrible victims and sufferings of millions in Russia, China, and Kampuchea.

Our contemporary Yuri Zhukov (a historian and inspired singer of Soviet socialism, see his article in the magazine "New world," №7, 1978) also called history "myth-making." He tried to imagine the monstrous atrocities of Stalin and the millions of victims of his regime as the myth. Instead, he created another myth: Stalin was clever, soft, kind, and intelligent, but only Communist Party officials did not allow him to be noble and gentle towards his people. And these stories are called history…

History is not science; it is the property of kings, emperors, kings, secretaries general, and presidents. History is elusive even in the minds of the best historians. Nikolay Karamzin—the great Columbus of Russian history—proclaimed that the historian was the designer of the past time. His task was to resurrect the past tense rather than to invent, to imagine, or to compose. But even his book "History" has appeared to be the basis for the well-known epigram written by great Russian poet Pushkin:

In his "History" elegance, simplicity
Prove to us without any addiction
The need for autocracy,
And pleasures of the whip.

Personal sympathy of the historian Nikolay Karamzin as an apologist of autocracy managed to emerge in his "History."

"History belongs to the peoples"—so Nikita Muravyov responded to the publication of three volumes of "History" in 1818; so he expressed his protest against its monarchical concept. However, Nikolay Karamzin had his own truth. He saw in history and in its content a reasonable beginning, embodied in the great people governing States. "Excellent minds are the real heroes of history"—Nikolay Karamzin wrote in 1802. And it echoes with the phrase of Rennan: "the goal of humanity is to create great people."

The great minds that control the States? It is something mythical; rather sarcastic… After all, ambitious people are eager for power, but ambitions, as we know, are only unsubstantiated claims on non-existent dignity (Duc de La Rochefoucauld). Exceptions are rare and random. An American economist Frank Hyneman Knight admitted: the probability that the authorities will be decent people, close to the probability that a good man will get a position "supervisor on the plantation." John Dalberg-Acton, first Baron Acton, added: "any power corrupts, but absolute power corrupts absolutely." "The path to prosperity and wealth lies through the corridors of power" (Benjamin Disraeli, first Earl of Beaconsfield).

Karamzin wanted to believe (it was his early religion) in progress and in the meaning of history, but he saw another… So the late wise Nikolay Karamzin realized that history was governed not by reason, but by a conflict of interests: wealth and poverty, top and bottom, freedom and slavery, pleasures and suffering, truth and lies, nobility and meanness, life and death. Power is the only driver of history, and it often changes its master. And at the same time, Nikolay Karamzin noticed that the story is the divine concert, "which is referred to as a case of the inevitability of a blind destiny." However, this endless concert of bloody crimes of people, nations, and states hardly could be called divine. It's like to hang up the call "Welcome" on the doors of the morgue. Both are inappropriate…

History as Myth-Making

Many historians treat historical events as their own property, accustomed to thinking that they make history. With the same monumental arrogance TV program, "News" states: "We make news." A history professor writes in the academic journal "Questions of history": "Our history still cannot get out of the shock caused by the collapse of ideological myths and schemas where it was a prisoner." And then, in the following phrase, he writes: "… the primary task is to develop a new concept." It never occurs to him that the second statement contradicts the first and that he calls to create new myths and schemes. This is his calling as a historian, and he is not alone. A pitiable vocation… Myth-making. An honest story is not science, but a history dressed in cloths of ideology is pure pseudoscience…

The main thing that history teaches is that it teaches nothing. Its lessons are bloody, cruel, inhuman and are absolutely useless; they are understood, but not perceived or used. When the current scientist-historian writes an article "Ten fatal mistakes of Emperor Nicholas II," it reminds a joke: "If I would be so smart now, as my wife later…" Talented Russian historian Yuri Polyakov once joked: "Life leaves a lot of watermelon crusts on the way of history. That's why it's so easy for a historian to slip." But he also has written a serious sentences: "History is mobile as mercury, and historical documents are unshakable as granite."

French writer Alexander Dumas noted: "History is a nail, on which you can hang anything you want." And historians hang… Here is the history of the Roman Empire. There are 220 theories about the causes of its death. And here is quite a short history of the October revolution of 1917 and its causes: the abdication of the king, the arrival in Russia of famous Bolshevik passengers in a German sealed car, the hated war, German imperialism, who bought the Bolsheviks, their arrogance and cruelty. And dozens of others, among which there is such one: Russian Tsar Nicholas II should make Bolshevik leader Vladimir Lenin the head of his tsarist government. (About Lenin: half of humanity mourned his death, although it was necessary to mourn his birth.) But Russian philosopher Nikolay Berdyaev believed that the Russian writer Nicolay Gogol is guilty as he drew (and did it brilliantly) disgusting portraits of Russian landowners. Of course, all this is empty profundity in the subjunctive mood. As the story of arguing passengers who left the train at the station:

And everybody went his own way,
But the train went on its…

Historians are similar to these passengers, and the story is like a train...

Historians' favorite exercises are to squeeze historical events into the framework of concepts, schemes, points of view, projects, trajectories, theories... The histories of Russia are written using the concepts of "communal conciliarity," "peasant communism covered with a thin layer of Marxism," "religious identity," and so on. There are as many concepts as many historians. Nietzsche and Marx proceeded from different conceptions: strong personality, superman and collectivism, socialism; however, they both led to a common result: bloody communist-fascist dictatorships. Both Christianity and Islam has declared itself as a religion and become politics; Bolshevism and communism have positioned it as a politician but turned into a religion. "There is everywhere their (the historians) theories... Everyone is free to put their theory on the altar of science and you can tell when it will dissipate the smoke: the smoke as the smoke is no worse than yours, sir..." (R. Töpfer).

Revolution is the source of historians' inspired creativity; everyone has his own vision, his own opinion, and revelation, his own search for causes and regularities. For some people, it was a monumental event that renews the world, but for others, it was a bloody abomination. The same is true for the revolutionaries: for some people they are heroes, others think them, villains. Well-known Soviet historian, academician Mikhail Pokrovsky graciously called the legendary Russian revolutionary Pyotr Tkachev the first conviction and the importance of a Bolshevik. Tkachev believed that to update Russia it is necessary to destroy all its inhabitants over 25 years. (The idea is not new: even during the French revolution of 1848, the citizen Duvivier offered to destroy all the French over 30 years; and Tkachev became an alcoholic and ended his days in Paris, in the house of the insane). The Bolsheviks failed to execute the Tkachev's project in full, but the proletarian leader Lenin demanded to publish all the Tkachev's nonsense. Not everyone knows that during the terrible famine in the Volga region in 1891, a young (he was 21 years old) assistant attorney Vladimir Ulyanov (then he became Lenin) vehemently objected to helping the hungry.

Lord Dufferin, the English Ambassador to the Russian court, gave his definition: revolution is a state where the killers are at the bottom, and suicides are at the top. Napoleon was the great woodman of Europe; he put in the ground for almost a quarter of Europe. He was the ambitious genius and savage in the service of civilization. Napoleon gave a more precise definition of revolution: it is the idea that got itself bayonets. The struggle of opinions around revolutions is a fruitless exercise; revolution is an unpredictable phenomenon with critical parameters. Such phenomena in physics are well known; the brightest of them is an earthquake. This is a nonlinear phenomenon with feedback of control parameters when an infinitesimal change of any parameter can produce a catastrophe. Prediction of earthquakes and revolutions are non-scientific virtual exercises. Both may or may not happen. And you can't predict when it will happen. Russian poet Alexander Pushkin wrote about revolutions and riots: "Only education is able to prevent the new follies and new social disaster." The result of revolutions is nearly always unambiguous: "the people killed their freedom…" (Pushkin). As a nonlinear system, history is the realm of chance.

The Power and the History

History is politics for all authorities; it is too important to give it to historians. It is kept in the archives and secret vaults; it is strictly guarded. Authorities always chased or just destroyed honest historians. But there are a lot of historians who willingly execute the order of the authorities wishing to decorate its acts and even to fix definitions of historical events legislatively. This property was called a new sense of history. Mark Aldanov summarized it in the law of history: the force always becomes truth. It becomes assigned truth, in which crimes are considered heroic feats, meanness is considered virtues, and the destruction of people is considered the boon. The official history is dominated by ostentatious boast of false power, even when it is followed by exhaustion and terrible poverty of the people, which has become customary…

The abovementioned Mikhail Pokrovsky was one of the most notorious and aggressive servants of authorities. He put forward the principle: "History is policy overturned in the past." A new story, which is created now by Russian academician Anatoly Fomenko, is no more than a brilliant parody of the Soviet state historians. Those historians were the prophets of official fake history, who wrote by order of the authorities and stood at the trough of authority.

In the ninth volume of the "History of Russian State" published in 1821, Nikolay Karamzin described the cruel reign of Ivan the Terrible, who was "the giant of inhumanity." But the priest Filaret (Metropolitan of the Russian Orthodox Church) advised to the author: "…it would be better to highlight the best part of the reign of Ivan the Terrible and to leave the other part in the shadow." The wishes of both secular and church authorities delightfully coincide: both want to be beautiful, both want to drive the pen of historians, both discover vain and restless claims to greatness. Nikolay Karamzin could be mistaken, but he was not a servant of authority; he had the strength and dignity to say to the mighty tsar: "I don't need your favors." This lesson is not suitable for all historians…

In the revolution of 1917, the Bolsheviks declared Napoleon the liberator of Russia for the fact that he fought with hated Imperial Russia. And people often have to wait a long time until the "year of hard times will smell the truth" (poet Igor Guberman). And could we say that history cannot be reversed? The possible reply is: it could be possible if it pleases the authorities... Even the gods cannot change the past, but historians (and not only Russian) can...

The current reforms of education in Russia almost reproduce those that were in 1850 in France resting from the bloody Napoleonic wars. Then professional education was considered unnecessary and teachers got a "certificate of obedience" signed by the local village priest instead of a diploma. The reforms were abolished in 1863 when another Minister came, but the damage was enormous: science in France was pushed back half a century.

History of Unfulfilled

Yes, there is such a paradoxical history... Historians have invented a very accurate statement: history does not tolerate the subjunctive mood. In fact, however, they ignore this statement completely and create the "science of hindsight" (remember: if I was so clever...). It is such an exciting topic! Even Napoleon thought, being on the island of St. Helena: "maybe it would be better for mankind if neither I nor Rousseau had ever existed." Maybe... However, what about Jean-Jacques Rousseau? His book "On the Social Contract; or Principles of Political Rights" was the prologue of the bloody revolution in France.

This is a wonderful exercise—to compose unfulfilled history: what would have happened if... Dozens of books and hundreds of articles on multivariate, alternative, and virtual history are published. However, there is no more than boundless fantasy, fascinating and fruitless. And it has nothing common with science. History, like life, consists of irreparable.

It is true that Russia is a country with a shaky and changeable history, the country with an unpredictable past. And it is no exception... History in the hands of interpreters is similar to the kingdom of crooked mirrors. In this sense, historical science is not far away from pseudoscience, or, to say it mildly, from parasciences that are products of spell casters, soothsayers, psychics, and so on. The concept of "official" is often applied to history, implying that there is also an informal history. But this does not mean that the second is better...

History Is the Great Culture

History is a river of events with spills, twists, and floods.

> *Boiling water of history*
> *Sweeps away all obstacles and dams.*
> (I. Huberman)

Historian is an archaeologist of human civilization of bygone eras. Only his finds have a price. Any attempt to interpret and to search the reasons (especially to find them!) is already a lie. There are no truth tests. When historian interpretations are offered, it provokes answer with the words of great Russian poet Mikhail Lermontov:

> *If you're offering me your soul,*
> *I want to damn both you and soul…*

Historian of culture Yuri Lotman noted that the work of an honest artist had contributed more to the progress of mankind than the preaching of moralist and interpreter. History is a chronicle of events, actions, and destinies of people, nations, and states that are participants of the monumental historical performance, which had no script. History as a sequence of events and destinies is free and independent. All Russian historians recognized as great (Nikolay Karamzin, Sergey Soloviev, Nikolay Kostomarov, Eugene Tarle, Yuri Lotman, etc.) were just talented writers and chroniclers. Those they were and continue to remain interesting, readable, and revered. We should not forget the grandiose historical works of A. I. Solzhenitsyn on the investigation of Communist-Bolshevik crimes. But there is no historical science… Alexander Nemirovsky, a modern brilliant historian, rightly noted that history should be presented no worse than belletristic literature, but without fiction. The story is interesting when it "extracts the raisin of amusement from the unleavened bread of history" (Alexander Dumas). Konstantin Simonov, a wonderful poet, and writer said something else about the history: "you cannot pick out of it only raisins like a five-year-old child from a loaf." And both were right… Russian historian Sigurd Schmidt noted that foundations of history are source studies and historiography.

The history as a science is deceit or ignorance interpretations, false concepts, inventions biased minds, the falseness of composed patterns. The place of all this is on a garbage heap of history. The only documentary should be left as pure history in documents. Both history and literature is not more than a mirror of life: history is a mirror of events; literature is a mirror of characters and souls. Both are able to teach a little. A person is taught only by personal experience. Someone else's experience does not touch or quickly forgotten and therefore is useless.

As a graveyard of someone experience, both history and literature have a low price. Dostoevsky's heroes have been ruthlessly condemned, but their number did not become less... Fascism has been condemned at Nuremberg, but he'll be back. The crimes of communism and Bolshevism are not condemned, and will also return... It is unknown only when and where... The present does not need the past, and the future does not accept victims from the present...

History is a biography of people, nations and states, a mighty river of events and biographies. Scientific problems of history might be far-fetched only. History is similar to Darwinian evolution: both do nothing on purpose and always are right. But if Darwinism is the history of biological evolution, then human history is only a special case of biological evolution.

Philosophy

Philosophy as a Theology of Civilization

> *The misfortune of philosophy is that its opinion about the world pretending to be knowledge. Philosophy is not a science; it is a state of mind and style of thinking...*

Sciences, like people, are born, grow old, and die. That's what happened to philosophy: it has died as fundamental science. It was the first and oldest science and was born together with humanity from properties of mind to watch and to ruminate. For centuries it was a science about the world and about nature, it was natural philosophy. The rapid progress of mechanics, physics, mathematics, chemistry, and molecular biology deprived the natural philosophy of its main subject that is nature. Science has taken this subject. Abstract concepts and speculative reasoning were replaced by precise knowledge. Science supplanted the philosophy, taking over all its concerns, and then "the death-hour of the old metaphysics came, and the fresh breath of nature appeared in the cells of abstract thought" (Friedrich Engels). Science has developed its own search technology, the logic of building hypotheses, analysis, and control, verification of consequences, and methods of testing their strength and reliability. It has created what is the great value of civilization and what is called the scientific method.

Natural philosophy as science is dead. Philosophical faculties, magazines, specialty, the title "doctor of philosophy" (which is irrelevant to philosophy),

© Springer Nature Singapore Pte Ltd. 2020
A. L. Buchachenko, *The Beauty and Fascination of Science*,
https://doi.org/10.1007/978-981-15-2592-6_9

etc. are relics of natural philosophy. It seems strange that a poorly educated student, who graduated from the faculty of philosophy, get loud (albeit useless) profession "philosopher." Neither great thinker Albert Einstein, nor his colleagues called him a philosopher. As the scientific journalist Alexander Mekhanik noted, today there are philosophers, but there is no philosophy. All real scientists of all sciences were real, profound philosophers, although they did not profess philosophy as a science; moreover, they were negligent to it…

Having lost the object, philosophy became helpless, lost its charm, its attractiveness and elitism. It fell apart. Its new status was firmly defined by Richard Chace Tolman: "Philosophy is a systematic confusion of terms specially invented for this purpose." However, the wise Einstein protected philosophy. He brilliantly recognizes the new status of philosophy: "Philosophy is like a mother who gave birth and put all other sciences on its feet. So it should not be despised in nakedness and poverty." It could be possible to agree with the latest if philosophy was not so jealous and aggressive, so fiercely hostile to those children, to whole science if it would not was so arrogant and ambitious. (Recall: ambitions are unfounded claims on non-existent dignity.)

Philosophy in the Opinions of People

The great Newton was a theologian and philosopher. He discussed the prophet Daniel, interpreted religious dogmas and the Apocalypse. He published the great book "Mathematical principles of natural philosophy." However, there was no philosophy, but there was mechanics that is the foundation of classical physics. The circumstances that accompanied the publication of this book forced Newton to write in a letter to Edmond Halley the following words: "Philosophy is such a brazen and vexatious lady that to deal with her is like being in litigation." Newton-philosopher did not favor philosophy. But Newton as the philosopher and Newton as the physicist are incommensurable: how insignificant is the first, so great is the second.

Russian philosopher and priest Pavel Florensky (a mathematician by education, one of the idols of philosophy) was superstitious and hypochondriac. Lydia (the daughter of the poet Vyacheslav Ivanov) remembers that 1 day on a walk with a Pavel Florensky she found a pin and wanted to pick it up. But frightened philosopher warned her: "Leave it. Sharp things are dangerous, they can be enchanted." Hence, perhaps, his fruitless religious mysticism presented as the great revelation of the spirit. Philosophy went close to and often together with mysticism, theosophy, occultism, Kabbalah, Shambala, and other garbage of consciousness (Elena Blavatskaya, Nikolay Roerich, Karl Bart, etc.).

Another Russian philosopher Pyotr Chaadaev highly valued science as he thought it "the supreme ruler of our century." Herewith he believed the law of gravity is a mathematical fiction and wanted to reduce all sciences in a strange way to Christian philosophy. Christian philosophy was thought as the perpetual and therefore trite-uncertain thinking about the meaning and image of "pathetic life," about the despair and anguish. French catholic priest and philosopher De La Mennais called "a special kind of idiocy." Pyotr Chaadaev's "Philosophical letters" is a big complaint that the world is arranged badly, not as he would like, not according to his ideas; his ultimate goal was the Kingdom of God on Earth. The goal is beautiful…

Auguste Comte, a French philosopher, was almost unknown during his lifetime, but became popular at the end of life. He tried to make philosophy an exact science. It was worthy attempt, but unsuccessful: in exact science there is a subject, in philosophy it is not present. Imre Lakatos is an amazing and popular philosopher; his "The Methodology of Scientific Research Programs" is evidence of monstrous ignorance, illiteracy, and incompetence; and they are proportional to aggressiveness. John Polkinghorne, a physicist, theologian and philosopher, criticizing Lakatos philosophy, admits that his "ideas are of limited value to theologians," i.e., they are so ignorant that they are unsuitable even for theology.

Here are the strokes to the portraits of some idols of philosophy. Aristotle is a great philosopher; all scholasticism came out of his teachings, in the center of which is the world created by the mind and reflections, the world is mythical and abstract, the world is fictional, unrealistic. It became a pillar of the Church and Church teachings with their aggressiveness and obscurantism. The Aristotelian philosophical system as a *reductio ad absurdum* of the ideological heirs of Aristotle, hundreds of years have frozen scientific and educational progress and European civilization. Galileo was the first to oppose a scientific system to empty scholasticism. Galileo's scientific system is based on a simple and clear thought: the source of knowledge is nature, its objects, and phenomena, rather than the scholasticism of the mind. Galileo paid a high price for this idea…

That is the great Friedrich Hegel (his contemporaries called him wooden), who was appointed the father of dialectics. His legendary statements "Everything really is reasonable. All reasonable is really" scratched on the tablets of philosophy. Both are true and both are false. They can put any meaning; it depends on the taste of the one who invests. And it means that this doctrine as a dialectical formula is wretched. This is the case when the idol is empty and the king is naked… It is known that on his deathbed Hegel admitted that there was only one person all over the world who understood him (bearing in mind, of course, them). And before his last breath said: "And, indeed, he didn't understand me." Hegel's greatness is not in his confused systems and concepts; it is in another—he taught people to think, to play with the mind. Here's Hegel: "in every science as much science as a philosophy." It is also nonsense. Immanuel Kant believed otherwise: "In every science as much science as mathematics." Though it is wrong, it is not meaningless. But the same Kant was capable of meaningless things: "things are not given to us in the real experiment; we know them through the influence on us…" The idol of the Russian philosophy Aleksei Losev said: "…the myth is the highest form of reality." Immanuel Kant, a great philosopher believed that space and time are concepts introduced into nature by the human mind. Hegel refused Newton that the white color is composed of colored components. At the request of the correspondent to express his philosophy, briefly, popular and French Hegel said: "My philosophy could not be expressed briefly, popular, and French." As you can see, the giants of philosophy did not bother to care about competence. That is a strange and ridiculous position of Nietzsche: "Not the victory of science is a distinguishing feature of our nineteenth century, but the victory of the scientific method over science."

Philosophy is admirable: it has created the illusion of science, and banal, capital truths learned to pass off as wise revelations. Indeed, the cunning of thinking is that in any empty thing, in any primitive concept, you can put a deep meaning and build a deep, albeit false, philosophy. "Don't touch idols because gold-plating may remain on hands." Philosophy is full of such idols with fake or fragile gilding. Here is legendary Russian philosopher Nikolay Berdyaev. On December 8, 1939, when he went down to breakfast, he told his wife the following: "Now I for the first time precisely formulated a difference between my philosophy and others, i.e. the majority of philosophies. In other philosophies, the thought, coming out of the concrete, rises to the abstract to reach the general, universal and there to look for the truth." I start with the fact that is immersed in specific and, towering over it, looking for a shared, enclosed in concrete, and in total I look for truth. "Do you understand the difference, reader?" His book "Origins and Meaning of Russian Communism" is bright and talented, but the content and ideas of this work are ambiguous. Einstein, reading a thesis on philosophy sent to him for review, said to his friend Paul Ehrenfest: "I feel as if I need to swallow something without having anything in my mouth." The great Paul Dirac, one of the protagonists of twentieth-century physics, was a student of philosophy, but quickly realized that "it is only a kind of fruitless reflections on the discoveries that have already been made" (Conversations with Dirac in Trieste, June 1968). Brilliant conductor and musician Rudolf Barshay said that the student went to classes in philosophy (did not want, but to pass exams), read philosophers and then he had a question: if the wise philosophers have long understood how to live, why the world is so terrible? Russian writer Vladimir Nabokov noted: "Philosophy helps mediocrity to respect them."

Idol of Russian philosophy NikolayFedorov (1828–1903) was an uneducated and illiterate man whose philosophy is a ridiculous mixture of everyday banality and mysticism. He became famous for the ridiculous theory of the resurrection of dead ancestors. There is all a mystic, although Berdyaev somehow believed Nicolay Fedorov was enemy of all mysticism.

Encyclopedia of Brockhaus and Efron wrote about Berdyaev: "Berdyaev's philosophical articles are not scientific research, which would have to be a philosopher nor a living essay, and interesting for contemporaries. They are heavy, wordy, contradictory, meaningless and confusing" (remember his story to his wife at breakfast). Instead of the mind, there are philosophizing. He is considered a theologian for he talked much and fruitlessly about religion. In his books, there are turbulent flows of empty words, vague and contradictory views, retellings of Kant, Nietzsche, Solovyov, etc. In Brockhaus and Efron's encyclopedia, he was called a helpless philosopher. The famous philosopher Ludwig Wittgenstein (1889–1951) admitted that all philosophical reflections are meaningless, including his own. Napoleon, a historically ambiguous figure, considered both the crooked Pope Pius VII and Kant charlatans.

Everybody should be careful dealing with philosophy. Excessive trust in it attempts to live according to its instructions; thoughtless passion for it is as dangerous as excessive alcohol consumption. Here is an example of Russian literary critic Vissarion Belinsky. He rushed about in life, getting acquainted with different philosophical doctrines, naively trusting and following them. His life turned into philosophical torture. He believed Hegel (all that is really reasonable...), opened a new world in him and accepted a comforting philosophy: found good, happiness, harmony, and the highest state of mind in the apologetics of slavery, serfdom, and tyranny. However, later he came to the realization of "vile reality," but this philosophical catastrophe left a bad mark in his life. And it is not only the fate of the mutilated philosophy. Reading philosophers almost brought Russian writer Gorky to mental illness (he has a story "about the dangers of philosophy"). The tsarist Minister of education Platon Shirinsky-Shikhmatov (1790–1853), who replaced Sergey Uvarov as Minister, was not so wrong when he banned the teaching of philosophy on the grounds that "the benefits of philosophy had not been proved, but the harm from it was possible." Note that in the Soviet (and in modern

Russian) historiography both Shirinsky-Shikhmatov and Uvarov are depicted as monstrous reactionists and stranglers of science and education. However, the Uvarov's Ministry was the wisest patron of the Moscow University and under Shirinsky-Shikhmatov's direction, University achieved prosperity and complete freedom. Minister Platon Shirinsky-Shikhmatov was a writer, archaeography, and academician of the Russian Imperial Academy.

But Alexander Herzen (unlike Belinsky) for some reason immediately found the algebra of revolution in the philosophy of Hegel. And this is a delightful property of philosophical systems: in them, one finds the "what is hunting." There are, of course, and the monstrous philosophical system (Benedetto Croce, Christian August Crusius) that are like the raving of people with unhealthy minds.

Unaesthetic Thoughts on Philosophy

Philosophy looks quite respectable and honorable activity. And it is almost science, when it talks about the spirit and the vague soul, about good and evil, about death and immortality, about space (but this is not the space that Gagarin and Armstrong visited), about the Apocalypse and little-known chiliasm, about life after death, about religion and God, about the national mind. But invading science it reveals helplessness and insignificance. Knowledge compiled from encyclopedias is not yet encyclopedic knowledge. The profundity of the philosophy is deceptive, there is emptiness behind it. Einstein: "doesn't philosophy seem to be written in honey? When you look at, it looks great. But when you look closely, there is only a mess of sentimental nonsense." Heinrich Heine responded mockingly about Hegel's system, which claims to reconcile the irreconcilable with the help of several formal postulates:

> *He darns gaps in the Universe*
> *With old dressing gown and other rags*

Famous German play writer Berthold Brecht considered Hegel's work "Science of Logic" as humorous literature that talks about a lifestyle of concepts that are ambiguous, unstable, irresponsible creatures that are always swearing: as soon as the Order says something, the Mess immediately refutes it. The great legendary physicist Richard Feynman, Nobel laureate, remembered how he once got to a seminar of philosophers, where they discussed the concept of an essential object. And humor it announced which a fierce argument broke out around the throne of the question: "Is a brick an essential object?" (R. Feynman, "Are you kidding, Mr. Feynman?").

Philosophy is immersed in reasoning about the known; therefore one cannot expect its new discoveries. It invents own concepts and so they are fragile and unstable. Immanuel Kant has transferred the concept of freedom from philosophical to legal one under the influence of the French revolution. But Hegel generally excluded the concept of freedom from philosophy as a term realizing the social danger of "universal equality." "The great philosopher" Vladimir Lenin defined the classical German philosophy as a theoretical source of Marxism. The totalitarian Communist-Bolshevik bloody finale of this source is known... Namely, philosophers are the creators of different social theories and religions, which disappear after turning into political doctrine. Sometimes with a roar, as it happened with the Communist ideology that was a philosophy for cooks...

In philosophical researches, people tend to magnify what they do not understand; this was noticed by Einstein in a letter to his friend Solovin. But the idea that there was nothing to understand came later. Or does not come at all... Russian play writer Anton Chekhov admired the talent of philosophers to pour from empty to empty with dignity. He also noted: "if someone philosophizes, it means that he does not understand." The verdict of Jean-Jacques Rousseau was: "To deny what is, to explain what is not is a mania of philosophers of all ages." The great André-Marie Ampère (the unit of electric current has got his name) noting the turbidity and inconsistency of philosophical concepts said (after Newton): "Philosophy—it is such a litigator." Eugene Delacroix did not like philosophy, considering it an empty and idle activity. Russian historian Vasily Klyuchevsky noted: "Philosophers peer into the depths of the sea of life and see there only their own faces." A brilliant mathematician Vladimir Arnold: "the philosophers are the most ignorant people in the world." And they are proud ignorance, they are proud of themselves.

There is no question that philosophy can answer. Its questions are doomed to be eternal. Different philosophical schools and groups give different answers even on their own invented questions. And every philosophical school and philosophical group fights for its own problems rejecting and humiliating others. Vladimir Vernadsky, a great thinker, and a bright man, never considered philosophy as a science. He portrayed philosophical schools and theories (rather not aesthetically) as columns along the road, which are marked by each running dog. Every great philosopher has his own philosophical theory, which he considers "the truest." Soviet philosopher Teodor Oizerman noted that it is the "personal nature of philosophy." But the personal character is incompatible with science…

Science and Philosophy: Who Needs Each Other?

Philosophers are convinced of the following. Philosophical intelligence is the truth of higher order; it is a mega-mind of a science. Philosophical truths, reflections, and problems are not available to the scientific mind. And to mind at all… And even more: "the philosophical mind, even if it is silent, still broadcasts…" (Vladimir Varavva, modern and fashionable philosopher). It sounds proudly, though ridiculous. Fascinating world of mirages and illusions…

Among the philosophers is popular proudly and indulgent formula: philosophy cannot be taught. This is true, because there is no subject, nothing to teach. And because philosophy is not a science, it is a style of thinking…

Respected by the philosophical community (and not only them) Merab Mamardashvili (1930–1990) considered philosophy as a pulsating continuum of consciousness, a means of honing and sharpening thoughts. But thought should have the subject. The wise Confucius taught: "To study something and not think about learned—it is useless. It is dangerous to think about something without studying the subject of reflection." Mamardashvili taught his colleagues: claiming for philosophical legislation in science, it is necessary to know science. And not from memorized phrases from school textbooks. And not according to encyclopedias: knowledge extracted from encyclopedias is not an encyclopedic knowledge… Extensive lectures of Mamardashvili is very boring and almost empty, although he was the amazing man. The more you study philosophical books and articles, the more convinced you are: philosophers do not interpret life (and Life) but grind it in a stupa… It is as fruitless as groundwater in the same stupa. By the way, the smartest of them realize it.

Philosophy is dangerous for its boastful claims of superiority over science (S. Kutateladze, "Philosophy as a utopia for culture"). The great mathematician Vladimir Arnold spoke about the corrupting influence of philosophy on science: "Science is often replaced by philosophical chatter and it is done by people who do not know how to do anything else." However, Marat, the idol of the French revolution, said earlier almost the same few moments before he was struck by the dagger of Charlotte Corday. It's tough, but it's fair. In fact, it is almost a Manifesto: natural science and natural scientists are not against philosophy and philosophers; they are against the ignorant philosophy and ignorant philosophers, aspiring to the role of the patriarchs in science. And it is not harmless for young minds and, therefore, for science.

Wise Montaigne 400 years ago noticed that simple peasants are beautiful people and philosophers are beautiful people, but all the troubles are due to semi-education. Vladimir Vernadsky noticed that philosophical statements can neither be proved nor be refuted. For the same reason, the young mathematician Maxim Kontsevich (he recently received the highest international award for theoretical physics exceeding that of the Nobel Prize) called the philosophy "slippery subject," where the correct definition and the right questions are impossible. No content, but the abyss of charm...

Almost all great scientists consider the philosophy condescendingly (sometimes hesitate to admit it) as if it is fruitless activity. After all, philosophy cannot say anything more than what science says. It is rightly noted that there can be no progress in philosophy because there is no criterion for the correctness of the results of metaphysical and abstract thinking. In philosophy, there is no logic of evidence. In everyday life, all barren philosophizing is called philosophy (often with humor, sometimes with irritation).

Among philosophers there are many intelligent and talented people who are aware of the poverty of philosophy. They came up with the slogans of self-affirmation: "Man, not armed with philosophy is a powerless creature"; "the totality of all philosophical foundations is an important part of physics"; "physics must be meaningful with the help of philosophical systems," etc. If we distract from the cunning of these statements, we must directly say that all this is senseless. Stephen Weinberg (Nobel Prize in physics) said bluntly: "I do not know any working scientist who takes philosophers seriously." Science does not need philosophy. Even the universal Russian idol of science Mikhail Lomonosov noticed with some irritation: "it is easy to be philosophers having learned by heart three words: "God created so," giving it in reply instead of all reasons."

Philosophers are inconsistent: trying to convince everyone of the superiority of philosophy, and they eagerly pounce on the latest achievements of science, adapting them and producing secondary products that are concepts, false interpretations, and false ideas distorting the essence of physical phenomena and theories due to ignorance and narrow-mindedness. So it was in former the Soviet Union with quantum mechanics, with the Darwinian theory of evolution, with genetics and molecular biology, and with the theory of relativity. Here even Einstein with his great tolerance and boundless generosity was forced to cut off: "...in my theory, there is nothing philosophical."

Where is the philosophy necessary? What is it for? Clever philosophers point out that philosophy is necessary for education, enlightenment, a culture of thinking, that its purpose is to form people with the best way of thinking. It is unlikely… Here the contribution of philosophy is less than insignificant. After all, the educational role of art and literature is negligible. Russian writer Varlam Shalamov noted: "I don't believe in literature and in its possibility of correction of the person." Dostoevsky was great and revered, but his hideous heroes were multiplied and triumphantly marched through life.

They say that philosophy is needed as a need of the human spirit, its aspiration to the high. This may be true, although it sounds like pathetic rhetoric. And it has nothing to do with science. And in the noble cause of education, enlightenment and culture of thinking the primacy belong to the Natural Sciences. Only science brings up a scientific view that is intelligent and sharp, attentive and attentive, incredulous and skeptical, kind and fair, looking forward and looking back, separating the main from the secondary, able to change their places, capable of analysis and logical synthesis… A mind is a divine gift of evolution…

In science, there is a global, universal and strict criterion of truth in its discoveries and ideas that is the principle of reductionism. He claims that the fundamental laws written down on the lower floors of the building of science are executed on the top. New knowledge includes the old as an integral part. There is no such reliable criterion in philosophy. Philosophers know the wise principle of their fellow philosopher William Occam, known as the "Occam's razor": "*Entia non sunt multiplicanda praeter necessitatem*" (no need to multiply entities unnecessarily). Newton, as a physicist, formulated this principle in his own way: "The rule 1. Should not accept in nature other causes beyond those that are true and sufficient to explain the phenomena." But it seems that philosophers do not follow Occam; on the contrary, they invent artificial concepts (a huge list of them can be found, for example, in Zenkovsky's books), extraordinary terminology, empty and false-meaningful concepts in which the mind is replaced by the philosophizing. Everyone piles up their theories, their concepts, which are dominated by rhetoric, the haze of double thoughts, contradictory mess of senselessness, verbal husk of emptiness. From a vicious web of thoughts, they weave concepts and images. There are a lot of banal pathos, empty pathetic and false ideas in their articles and books. They criticize a lot, not listening, not hearing, not understanding, and not wanting to understand each other. All this is accompanied by bombast of style and language. Academician G. Zavarzin noted that in philosophical books as well as in articles attention is focused on statements, not on judgments and factual generalizations. He as a natural scientist was puzzled why philosophers boil down everything to discussing the question of who said what. One of the pillars of Austrian philosophy, the famous psychologist Sigmund Freud, noted the main feature of philosophy is speculative thinking and generating useless hypotheses.

The tendency of philosophers to ambiguous things is well known. Soul, the mystery of consciousness, immortality, a transcendental subject matter beyond the limits of science and knowledge, or pseudo, simulating science "mysteries of the Bermuda" are the favorite dance floor of philosophers. The result of the philosophical battles formulated by the Swiss writer and educator Rodolphe Töpffer: "Everywhere they have theories, but there are no truths; many workers, but there are no knowledgeable. Everyone is free to put their theory on the altar of science and tell when the smoke will dissipate: the smoke as the smoke is no worse than yours, sir…" The fragility of belief, the vagueness, and uncertainty of the opinions, the instability of positions and systems are accompanied by a struggle, the literary wars, and often denunciations. The power skillfully uses both philosophy and philosophers. This was recognized by Arthur Schopenhauer: "Governments make philosophy a means of serving their interests." It is possible to remember how in the Soviet Union the ignorant philosophical pack was lowered on genetics, on cybernetics, on chemistry and the theory of relativity. The government has turned philosophy and philosophers into a political bludgeon and in ideological police, custom maid of political regimes that were killing science in the same way, as did the medieval Inquisition. Repressed and shot science… The campaign against science was led by Marxist-Leninist philosophy that was a miserable system of false values, which was imposed by the tyranny of power everywhere—from iron smelting to gynecology. Witty Alexander Minkin noticed that the delicate taste is incompatible with Marxism-Leninism philosophy. Vladimir Vernadsky riskily objected to the election of Abram Deborin as an academician of the USSR, arguing that the philosophy, which Deborin represented (and it was, of course, Marxist-Leninist philosophy), was not a science. To save himself and his relatives, the great physicist Vladimir Fock was forced to write a monstrous lie: "The philosophical meaning of our views on the theory of space, time and gravitation was formulated under the influence of the philosophy of dialectical materialism."

Several generations at all social levels were forced to learn all this philosophy. And now little has changed… Such violence gave rise to disgust the philosophy of almost everyone and especially the scientists—physicists, mathematicians, chemists, biologists. A brilliant Russian physicist Arkady Migdal: "teaching philosophy causes the same disgust in thinking people as forced feeding in children." The warmest and friendliest attitude to the philosopher is expressed in words: a philosopher is a person who talks a lot and fruitlessly (meaning, of course, a profession, not a person). Poet Boris Pasternak gave such philosophers a precise definition: "priests of the Marxist parish." The definition is offensive because in essence philosophers are good people, but they are vulnerable to the narrowness of their education. Students of natural faculties study philosophy by force, but students-philosophers study neither physics, nor chemistry, nor biology. Why? Professional arrogance? Intellectual powerlessness? And it is a mystery how it is possible to create the philosophy of science knowing nothing and not understanding the science?

Do not think that only the Soviet philosophy successfully strangled science. Philosophical agony continues to poison science now globally. Positivism as a philosophical doctrine states that the criterion of the correctness of the theory is an experimental test, and each element of the theory must be based on the observed values. This is a true, albeit very limited, a wretched point of view; its generalization is the evil that brings positivism to science and civilization. The great Ernst Mach, who remained forever in the physics of shock waves (Mach number), a brilliant mechanical scientist and a fierce positivist, completely ignored the successes of the atomic and molecular theory of the nineteenth century only due to the pathetic grounds that atoms and molecules were not observed directly. It was his authority that was the reason that modern atomistic as the fundamental basis of the Universe was recognized only in 1908 after Ostwald's statement on the origins of the Brownian motion. And now single atoms and molecules have been observed and work in optics and radio spectroscopy, in electron and atomic force microscopy. Einstein valued Mach highly as a scientist and teacher, but referred to him as a "pathetic philosopher." Albert Einstein (like most natural scientists) treated philosophy as religion and the theology of civilization. This was a wise position…

Positivism fights fiercely against quantum field theory and quantum mechanics, against quarks and gluons, against Big Bang theory and quantum chromodynamics. Universal world parameters (gravity constant and Planck constant that were the first swallows of the strange and mysterious quantum world, etc.) were considered by positivists as fiction. In their opinion, the laws of nature are invented and do not exist in reality. They ignore the fact that the consequences of these laws have got experimental confirmation and justification (including both theories of relativity). And a stronger confirmation than an atomic bomb, you cannot even think of… As Max Planck noted, for the philosopher-positivist stars are just complexes of optical feelings rather than real things. Philosophers-positivists believe that quantum strings are inventions that violate "Occam's razor." They do not want to understand that quantum strings is a necessity, that can eliminate the incompatibility of the two is absolutely correct and accurate theories—quantum mechanics and relativity theory. These people are so primitive that they do not want to understand: the brain and thinking are many orders of magnitude smarter and more insightful than the most observant look of eyes. Behind all this is the wild ignorance of philosophers-positivists. In Buddhism, ignorance is considered poison; Socrates took all evil as a consequence of ignorance. It is necessary to put obstacles in the way of blind progress rather than blind obstacles in the way of progress. The great philosopher Karl Popper teaches that any scientific theory that cannot be disproved is false. Another modern domestic philosopher persistently preaches that technocratic, scientific, and engineering thinking kills the spiritual world of man. All this, of course, nonsense…

Of course, positivism cannot destroy science, but its attacks against science are not harmless. It is capable to slow down the movement of science and depress the young talented minds. In addition, it disorients those who finance science (they usually do not suffer from excess knowledge). In fact, positivism means a ban on thought, on mind play, on imagination and inspired thinking…

The other dangerous (although ridiculous) philosophical doctrine is relativism. It declares that science is subjective and unable to reveal the objective truth and it is a kind of myth-making, religion, and shamanism. And it also comes from ignorance, from inability and unwillingness to understand science. This ignores and hypocritically ignores the fact that the reliability of new knowledge was tested repeatedly and comprehensively at all levels of knowledge. The doctrines of the "new philosophy" declare science alien and even hostile to the human spirit; they introduce apeiron, entelechy, transcendentation, eidos, hylozoism, panpsychism, and other concepts that are devoid of truth criteria. This obscurantism is covered by the name of philosophy. The famous philosopher Paul Karl Feyerabend put science in line with mythology and religion; he demanded to separate the state from science "this most aggressive and most dogmatic religious institution."

And about Russian philosophy. It was born late, so it was not a natural philosophy; science and the structure of the world were not its central themes. Russian philosopher Nikolay Berdyaev was right noticing that only three themes became the main subjects: religious, ethical, and social. And this tradition has been preserved: modern Russian philosophy has gone into political science, sociology, logic, history, moral problems, and economics. Here it has found its modest and worthy place. But funny and relict ambitions to teach physicists, chemists, and biologists are still being conserved. Here it is appropriate to recall the famous parable: "You should judge, my friend, not above the boot"…

Aesthetics of Science

Science Is a Leader of the Civilization

Outside of science or outside of modern science? Here is the intrigue: whether are there things unknowable and how can we distinguish them from unknown things? Is our capacity infinite for knowing the world? Are there things beyond science? However, in the last question two ones are implied: "out of science?" or "outside modern science?" The difference between them is the same as between the statements "my house is my fortress" and "my house is the Peter and Paul fortress." Here it is worth to note that the Peter and Paul fortress in Saint-Petersburg was a prison in nineteenth century. And this difference was acutely aware of the Russian Decembrists…

In Aristotle's time, electricity was outside of science. But under Faraday, it became an element of civilization; but lasers, radioactivity, computers, mobile phone, TV, and more remained outside of science; all these are the bases of modern civilization. Recently, a decade ago chemistry did not have the thought of a single molecule as an object of research and knowledge. And today it is perfectly mastered area of chemistry. And a transistor on a single molecule has been already created; there were single-molecule magnets and the real contours of a new technological civilization—molecular electronics, in which the functional elements are single molecules.

Even meteorites were outside science. The French Academy of Sciences in 1768 discussed the case of a meteorite fall and came to the conclusion: the stones from the sky cannot fall. The great chemists Antoine Laurent de Lavoisier who was appointed an expert on this event gave the following clever judgment: a meteorite is an ordinary stone melted by lightning. Talks about

© Springer Nature Singapore Pte Ltd. 2020
A. L. Buchachenko, *The Beauty and Fascination of Science*,
https://doi.org/10.1007/978-981-15-2592-6_10

the influence of the magnetic field on chemical reactions until recently were considered as a sign of shameful ignorance, it was beyond science. But today new fields of chemical physics have been established—spin chemistry, chemical radio physics, and chemical polarization of nuclei; they brought major discoveries of new magnetic isotope effects, magnetic catalysis and new magnetic phenomena. Here science has destroyed old dogmas and prejudices as the wreckage of the old truth, and has brought a new truth.

Physics, chemistry, and especially biology know a lot of such unexpected and almost magical transformations of phenomena and events from the state "outside of science" to the state "as it should be." A miracle is something that has no reason. That is why the search for the causes of "extra-scientific miracles" and the inclusion of miracles in a number of "legitimate" scientific discoveries is one of the great charms of science.

> *Everything was said.*
> *Nothing is untold.*
> *And unspeakable light*
> *Yet somewhere glows.*
> (Novella Matveeva, Russian poetess)

Namely this light of the unknown that calls people of science… They have a professional race for the new and the unknown. Mystery is always fascinating. "The most beautiful and profound experience that falls on the share of man is a sense of mystery. It is at the heart of all profound trends in science and art. Anyone who has not experienced this feeling seems to me, if not dead, then, in any case, blind." This was said by Einstein.

Scientific creativity is the transformation of the unpredictable into the inevitable. But is everything amenable to such a transformation? One of the charms of science is the temptation to find the answer to the most difficult question—why? Why are perfect theories so perfect? Why Euclidean geometry and Newtonian mechanics so accurately describe the macrocosm, and quantum mechanics describes the microcosm? Why are Maxwell's equations and relativity theory so accurate? Why does the world exist with such fundamental constants? Why does consciousness exist, where did it come from? Why in living organisms all proteins are built of "left" amino acids (rotating the plane of polarization of light clockwise), and all polysaccharides—of "right" molecules? Why do neurotransmitters synthesize in one place and work in another? Even if we understand how neurons and synapses work, that's not the answer to why. The electron doesn't fall on the nucleus, it holds the world… Why? Even quantum mechanics, which knows everything, does not know the answer…

At the turn of the nineteenth and twentieth centuries, the great Sigmund Freud wrote: "absolutely untenable is any attempt to determine the brain localization of mental processes and show that ideas, being present in the nerve cells and exciting them, move along the nerve fibers." Today, this statement has become a mere archaism. Ludwig Wittgenstein (1889–1951) in his work "on reliability" mentions someone who said that he had been on the moon. "It's just a joke; it's impossible to get up or fly there. All our physics forbids to believe in it." Such is the illusory nature of the indisputable provisions of common sense…

The origin of life as a phenomenon, the work of consciousness, and many other issues remain unresolved. Of course, you can dismiss them and say that they are inappropriate. And the eternal charm of science and calling the magic of mystery is the ability of science to give suddenly answers to mysterious questions. Or not to give them at all… The magic of knowledge carries hopes (although, according to wits, hope is just a delayed disappointment). Still not solved or not solved?—is open question and this question is not for philosophers. It is just a matter of science. Science of the future… Are there any questions you can't ask science? No such questions. Are there any unanswered questions? Yes. And it is the only question of God. Just because God is beyond science.

Who Makes Science

To learn, to discover, to inform –
It is the fate of the scientist.
Francois Arago.

Science is made by people… and they are not spared from bad manners: there is a place of envy and betrayal, ill will and careerism, stupidity and meanness. But there are also brilliant examples of great nobility, self-sacrifice, and aristocracy. The intellectual avant-garde geniuses inspire science. Archimedes, Richard Feynman, Robert Woodward, Dmitri Mendeleev, Heinrich Hertz, Niels Bohr, Johannes Kepler, Alexander Popov, Andre-Marie Ampere, Oliver Heaviside, Pyotr (Peter) Kapitsa, Max Born, Yakov Zeldovich, Conrad Lorenz, Gottfried Leibniz, Lev Landau, Ludwig Boltzmann, Michael Faraday, Nikolay Semenov, Jean Fourier, Carl Gauss, Nikolai Lobachevsky…

This is only a very modest number of names that are widely known and belonging to the world and history. Their discoveries changed people's minds. But there are names—Galileo, Newton, Maxwell, Darwin, Einstein—whose discoveries changed the consciousness of mankind. And there are hundreds and thousands of little-known or forgotten names. They were different—kind and arrogant, generous and greedy, seekers of glory and quiet workers. Some were bright, sparkling, others dull and boring. Somebody alive appointed himself as a genius; others, being really genius, shunned and ashamed of such recognition. Charles Darwin and Mendeleev belonged to these "others." Darwin wrote in his autobiography: "I have neither the quickness of mind nor the wit so prominent in some clever men," "What a genius," cried Mendeleev indignantly, when he was so called, "worked all his life, that's a genius…" All they were personality with their talents and discoveries, mistakes and blunders, with their ups and downs, with their wise focus and the strange absentmindedness. Real scientists are enthusiastic and modest, as Russian poet Pushkin wrote:

> *Deep water smoothly flow,*
> *Wise people quietly live…*

However all of them, regardless of their characters and personal attachments, did what is called the steps of civilization. They made major discoveries, gave rise to great creative ideas on which humanity built the floors of the majestic building of civilization. Isaac Newton was the absolute supreme genius, but he quarreled with almost all contemporary scientists. He 30 years cruelly argued with Leibniz for priority in the discovery of differential calculus and did not hide the joy when Leibniz died. Antoine Lavoisier (1743–1794), the founder of modern chemistry, was an arrogant and terribly conceited man. Jacobins sent him to the guillotine with the sentence: "the Republic does not need scientists."(The Bolsheviks expelled the intellectual elite out of the country from Russia in 1922.) Later, someone bitterly noticed that only 1 s was needed to decapitate Lavoisier, but to have the similar new head, you have to wait a century.

Charles Coulomb (1736–1806) was a brilliant experimental physicist. The law of interaction of electric charges has got his name—Coulomb's law. It defines the force of attraction or repulsion of charges as a function of the distance between them. Even schoolchildren know that this force is inversely proportional to the square of the distance. But they hardly know that the law is accurate to 10^{-16} cm and that it is not empirical, i.e., found experimentally, as Coulomb himself thought, but a fundamental law following exactly from

Maxwell's equations. During the French revolution and terror, Pendant was forced to leave science. Like other scientists, he was removed from all affairs with the wording: "not credible due to lack of revolutionary prowess and hatred of kings." Well, just like the Bolsheviks, who expelled the smartest people from Russia for the same reasons.

Well known in Russia biologist Kliment Timiryazev unjustly accused his teacher A. Famitsin of anti-Darwinism. During Communist Party leadership, the Timiryazev's political credibility was enough to make forgotten for half a century the name of Famitsin who was a brilliant scientist, an outstanding biologist, and the founder of the Russian plant physiology. Timiryazev also badly influenced the fate of another brilliant scientist, physicist Pyotr Lazarev, who was the creator of the molecular concept of photophysics and photochemistry. Pyotr Lazarev was arrested in 1931, his Institute of physics and biophysics was defeated, and his wife Olga Lazareva in despair committed suicide. And the further scientific fate of Pyotr Lazarev was under the pressure of the political regime.

Of course, you can reset something at the expense of the customs of those times and regimes. About them, Yuri Nagibin sadly said: "I'm afraid, my great idea to live a life until the end as a decent man will not work." But Pyotr Kapitsa boldly and firmly stood in defense of Lev Landau and saved him. The great Russian biochemist Alexey Bach signed (along with other scientists) a denunciation of a brilliant scientist, biologist, and geneticist Nikolay Koltsov and helped to ruin him. Moreover, academician Alexey Bach wrote the slanderous article in the Communist-Bolshevik newspaper "Pravda" about Koltsov that was titled "Pseudoscientist has no place in the Academy of Sciences of the USSR." And this meant a death sentence: Nikolay Koltsov died of a heart clot. The next day his wife and an assistant Maria Sadovnikova-Koltsov committed suicide…

Henri Poincaré, the greatest mathematician of the nineteenth century, rejected the theory of Ludwig Boltzmann, who was the great physicist, the Creator of nonequilibrium thermodynamics, and the author of the fundamental concept of entropy. Poincare did not recommend learning the works of Boltzmann because it was contrary to Poincare's ideas. However, finding a mistake in his already published article on the three-body problem in mechanics, Poincaré spent the prize awarded to him for this work to buy out all copies of the journal with this article and send out the corrected work to all subscribers. Many subtle observations and stories about scientists are in the excellent book by Y. A. Zolotov, "Making science. Who are they?".

In science, some events are known that surpass anecdotes. Not everyone knows that the Royal Society of London (the British Academy of Sciences) elected the courtier Alexander Menshikov as its member. Menshikov was friend of Russian Tsar Peter the Great; however, he was an illiterate man who could neither read nor write. A diploma of election was signed by Isaac Newton. There are idols and antiheros in science as well as in human society. Some love the process of studying and searching; it's their affection and the meaning of life. Others love themselves in science... As well as in life:

One needs a crown,
Others need love...

Scientists are very different... Some are striving for a career, for positions, and for titles. Having achieved them, they bring themselves everywhere as a gift. Even at the funeral they consider themselves superior to the deceased, and on anniversaries—the main hero of the day. For many scientists, a rank seems more delicious than a ham... Losing ranks and positions, they become invisible. As noted by Russian writer Yuri Nagibin, they are similar to the invisible man described in novel of Herbert Wells: to become visible and notable, they need to turn into blankets, hauberks and bandages of awards, medals, diplomas, titles, etc. Ambition, issued for destination, and ghostly merits are combined with a manic sense of grandeur and exclusivity. But Socrates four centuries before Christ followed the commandment: "Whoever wants to be the first between you, he will be your slave." As a great citizen of Athens, Socrates avoided supremacy and avoided honor.

But science is much more valued intelligence and talent. There are many people who are passionate about science, and who are tied to science. To take them from science "is equivalent to put them in a coffin without waiting for them to die." They are internally free, modest and with dignity do their fascinating scientific work. Science for them is a profession and vocation, duty and affection, passion and devotion, care and charm, inspiration and work.

Recognition in science is a rare, almost ephemeral thing, similar to a mirage. The first one who said that the speed of light is finite was René Antoine de Réaumur (around 1670), but the world realized this only 300 years later. The great Oliver Heaviside (the one who predicted the existence in the upper atmosphere of an ionized layer reflecting radio waves, and who opened the way of mankind to modern computer science) knew the great formula $E = mc^2$ even 15 years before Einstein. But few people know about it. Still, the scientific community is one of the most democratic. To all kinds of ranks, titles, and positions, scientists are without subservience; real merits are respected. And there is an understanding that the craving for leadership is not a sign of the mind, but a property of character. Contribution to science is determined not by position, but by talent, although the opposite point of view is much closer. In general, the scientific community maintains standards of intelligence and rules of moral purity. Exceptions happen; they are noticeable and attract attention. In a healthy scientific community, there is mutual assistance, mutual support, willingness to inspire, combined with friendly criticism. For science is a delicate thing, in it, as in art, you need attention, encouragement, inspiration. This idea was expressed by great Ahmed H. Zewail: "in our scientific case, the delight of discovery is the real fuel for takeoff, but that the flight was pleasant, you need to come to recognition." And earlier Caesar wisely said: "Better to listen to the reproach intelligent than the song of fools."

Idolization

Forever disregard the commandment of the Old Testament "do not make an idol to you on earth, because rust and moth destroy." The bad tradition to create idols is eternal. On the one hand, the idol forgets that between the great and the funny—a boundary invisible to him, but seen by others. And so idols often and forever are on the funny side. On the other hand, false idols disorient people claiming false targets and false values. Russian poet Boris Pasternak subtly remarked that "being famous is ugly," as well as being rich and all-powerful.

And in life, and in science there are idols and false idols; often this one thing and the same. The worst case is self-designation (Kazimir Malevich, Marc Chagall). As government all-powerful commissioners, they appointed avant-garde art, and they are geniuses prescribing admire them. Their destinies were difficult and worthy of sympathy. Intellectual Alexander Benois, a contemporary of Malevich, noted that his "Black Square" is an act of self-assertion, which has its own name abomination of desolation; this is a Manifesto of non-objectivity and emptiness, and the artistic skills of these idols at the level of sophomores of art schools. When a person does not own a brush, he is appointed a primitive artist and declared primitive art. The same applies to masterpieces. Focus on the masterpieces has always been questionable…

However, this is a matter of taste—artistic and intellectual. There is a discrepancy between the Natural Sciences and not the natural; the last one was a distinguished physicist called unnatural. A brilliant physicist, the legendary Yakov Zeldovich even proposed a criterion, an indicator of the student's creative abilities: the difference between the examination grades in natural and social sciences. The greater the difference, the more talented a student is.

Alexander Merzhanov, academician, Creator of the "solid" flame science wrote an excellent autobiographical book "It is better to be needed than free." Another position: "I always thought one of the vilest things to be useful and necessary..." Of course, that's the kind of posturing. But it is Charles Baudelaire who was a fashionable and popular poet of France and because he posturing to face, it was sincere: this was the nature of this man. Of course, the talent is amazing, but when it is annoying, Intrusive demonstrates and presents itself, it loses both charm and attractiveness.

The literary idol of all times and peoples is Fyodor Dostoevsky, unsurpassed master of images, language, and psychology. Vladimir Nabokov considered him a mediocrity, whose books are full of melodrama, platitudes, and fruitless moralizing; the latter is true, the heroes of Dostoevsky are not getting smaller. All literature seems to be a giant graveyard of human experience. Dostoevsky was "unpleasant and uninteresting" to the genius Mark Aldanov.

Varlam Shalamov—writer strict and long-suffering, sharply negatively (and reasonably!) belonged to the works of Alexei Tolstoy and Maxim Gorky, appointed idols and favorites of the public. Eugene Delacroix is considered a genius of a powerful and rapid palette. And here is another view: "four-fifths of Delacroix's works are pure nonsense, the rest are of dubious or suspicious dignity" (Pierre-Joseph Proudhon, the one who said: property is theft). Delacroix's contemporaries also considered him an artist of the morgue, plague, and cholera, and what he wrote was painting bad taste. But tastes change. Federico Fellini recognized genius creator of movies. But here is the professional opinion of Russian writer Yuri Nagibin, an intelligent and talented writer, a brilliant screenwriter: "Fellini is a shameful impotent because he has everything so liquid and strained. But no one dares to say it out loud..." Nagibin is trustworthy.

Picasso is a genius and an idol of millions... But here is the advice of one famous artist: "Look with a clear and meaningful eye—and you will see falseness and sophisticated ugliness." It is wisely said: tastes differ. Jean-Paul Marat... Hero and idol. But his biography and his fate, like his acts, are disgusting...

A mysterious idol is Sigmund Freud. His "scientific psychology" is a simple set of observations and claims to scientific there are inappropriate. Worst of all, it draws exaggerated attention to the dark sides of consciousness, awakens the dormant sides of consciousness, hidden in the shadows, and stimulates the excavation of the worst in the mind. Abuse of them has the danger to turn a normal person into a neurotic. Freud himself admitted that fantasies of his patients he turned out to be pseudo-scientific psychotheatre instead of a strict science—"the physics of the soul", as he wanted to. For some reason, everyone forgets that Freudianism was born from the psyche of unhealthy people. Vladimir Nabokov called the Freudian "libido" as nonsense. Einstein shunned Freud and maintained rare contacts with him only within the framework of secular politeness. But the life and personality of Sigmund Freud are worthy of respect and admiration; the suffering of the last years of his life deserves deep sympathy.

From Freudianism comes a disgusting novel by Suskind "Perfumer," which for some reason is appointed a brilliant work. Another example: appointed genius writer James Joyce and his novel "Ulysses"… Vladimir Nabokov, a man of fine aesthetic taste, mockingly called him a "divine creation of art." But Sergei Eisenstein (a Soviet film director and film theorist) said: "Joyce had gone as far as it may go in literature. And even further, beyond…" The main meaning is in the last words.

Count Leo Tolstoy… His greatness does not need characteristics; his thoughts are eternal and eternal… But they are simple, there is nothing wise in them and they visit everyone. But not all of them declare. But his theory of non-resistance to evil is bad and harmful; if it were followed, the world would be at the mercy of evil and violence. However, it is enough, and no theory… It is curious that the Church has preached the same, but the enemies, even imaginary, destroyed ruthlessly. Count deeply understood people and aware of the world, its scale, but it was a fierce anti-Darwinist. And it is strange… Tolstoy had a lot of followers and admirers; many of them later became disillusioned and parted with Tolstoy. Their motives were expressed by Mikhail Novoselov (1864–1938), who was historian and philologist, Church writer, graduate of Moscow University, and a close friend of the Count. He broke up with him, writing in a letter to him the following lines: "Your God is only your idea, which you have fancied and continue to fancy, turning it from side to side for two decades. You cannot get out of the vicious circle of your own self." But true said… At Astapovo station, before his death, Tolstoy said as if bequeathed: "Only one thing I suggest you remember that there's a divide of people in the world, and you're looking at one Leo."

What is this? To remind Kant: "have the courage to live with your mind…" and Gamzatov: "do not saddle other people's thoughts… Start your own."

Birth and Destiny of Discoveries

Many beautiful voices in science were not heard,
and heard without response, without echo…

Science lives almost independently of society; the vast majority of people are absolutely indifferent to it, to its successes and discoveries. People know that all the benefits can be purchased in the store. They are not interested in where it all came from and they do not think about science. In science public acknowledgement comes, as a rule, rarely and late, or even never. Real scientists are usually modest; the extraction of knowledge is a constant touch of mystery, they are majestic and respectful, and the real greatness is always modest. New ideas are hard to accept, reluctantly. And the reason for this John Keynes noticed with guile: "the difficulty lies not in the generation of new ideas, but in salvation from the old." The old has earned the trust, it is proven and reliable. And then the creators of new ideas have to wait and believe after Marina Tsvetaeva:

My poems, like precious wines,
Will come its turn…

And it comes… Often after the departure of the creators into eternity.

To escape from old ideas prevents more love… Great Ernst Rutherford who discovered the nucleus of the atom and created the planetary model of the atom was extremely reluctant to recommend an article by Niels Bohr in print, although Bohr had further developed the planetary model in this article. But Rutherford did not need it, he loved his native model. And he was alien to Bohr's idea of the free will of the electron, although it was a continuation of his, Rutherford's idea of the free will of the nucleus: it disintegrates when he wants and how he wants—throws an electron or helium nucleus. The Great George John Thompson, Lord Kelvin, treated both the Rutherford and Bohr models with hostility: he loved his model of electronic cake with raisins, i.e., with atomic nuclei.

Many persons have heard of Lobachevsky's geometry; few know that its authors were (independently) three: Lobachevsky, Bolya, and Gauss. The first two were published, cautious Gauss abstained. Lobachevsky was tormented by ridicule and mockery of his colleagues—mathematicians. Among them, the most furious mocking was the great Ostrogradsky (remember the Ostrogradsky–Gauss theorem?). Great recognition came much later, and it seems Lobachevsky did not wait.

The great Max Planck, the founder of quantum mechanics, was a man with a tragic fate. In 1909, he lost his wife; his eldest son, Carl, was killed at Verdun in World War I. His other son Erwin was executed as part of a plot against Hitler. Both daughters—Greta and Emma—died; he survived Nazi fascism and nearly died under bombs. Already in old age (was born on April 23,

1858, and lived 89 years), he remembered that he never managed to get recognition of any new statement. His idea of quanta was not accepted immediately and not victoriously. Louis de Broglie considered it a witty trick; Arnold Sommerfeld also interpreted it as a successful form of explanation, not physical reality. Planck's report on December 14, 1900, in which he first announced the quantum of energy and introduced the quantum $h = 6.5 \times 10^{-27}$ erg × second (then it became a world constant and it was called the Planck constant), did not cause any reaction. Quanta went unnoticed. The quantum theory of radiation, created by Planck in 1900, was completely ignored; there was not a single article, not a single book about it…

Much later, in 1914–1916, on the other side of the Earth, in the United States, another great physicist Robert Andrews Millikan, being a fierce opponent of quanta, measured the photo effects on alkali metals. That is, he watched as the electrons are knocked out of the metal when it is illuminated by light of different wavelengths. It was a precise and subtle experiment. To clean the surface of the metal from which light knocks electrons, was clean, Millikan invented and made a device for cutting metal in high vacuum. The result was amazing: Millikan got exactly the same value of the Planck constant h that Planck predicted. Of course, Millikan was shocked, discouraged…

And only then the recognition of this wild, but absolutely true idea of quanta began. And much later, Einstein gave an accurate assessment of the great thing that Planck did: "…in addition to the atomistic structure of matter, there is an atomistic structure of energy…" And this crowns the great ratio $E = mc^2$.

Planck ushered in a new era in physics. Quantum mechanics is based on his idea. His name was included in modern physics and astrophysics: Planck constant h, Planck length (10^{-33} cm), Planck time (10^{-43} s), Planck mass (10^{-5} g), Planck energy (10^{19} proton masses in the energy scale $E = mc^2$), Planck tension (10^{39} tons, it exists in quantum strings).

Movement of science is continuous and almost imperceptible; only scientific breakthroughs are visible. The temptation to make a discovery is great, but there is always a danger of error, false discovery. Often an astronomer opens a new star, which turns out to be a candle in the window of a neighbor opposite; this warning accompanies every serious scientist. Even great discoveries were not recognized immediately, but often with difficulty, overcoming resistance to rejection. In science, there are no "Royal roads," i.e., roads of universal conquest and subordination.

Einstein's theory was incomprehensible and supernatural, as it broke with the usual Euclidean geometry. Legendary-great Oliver Heaviside understood it at once but considered it senseless. The famous physicist Albert Michelson called it incompatible with common sense. That was true. Nikola Tesla ridiculed Einstein's ideas and his formula $E = mc^2$ was considered an illusion. The great Ernest Mach ignored the theory. The theory was defined as the extreme degree of madness, crazy trick, a shameful idiot child brain, and pseudoscience; Dean of the faculty of astronomy at the University of Chicago Thomas See has written about Einstein: he's just a mess, his theory is a delusion and sophistry, if it claims that gravity is not a force, but a property of space… But the most disgusting of all was the furious anti-Semite Philipp von Lenard, a German physicist with fascist views. He considered Einstein's theory as Jewish science. By the way, the Nobel Prize was awarded to Einstein not for the theory of relativity that was the greatest thing that Einstein did, but for the physics of the photoelectric effect. In 1930, a book was published in Germany, "one hundred professors prove that Einstein is wrong." It was about the theory of relativity. Upon knowing of this, Einstein said in surprise: "Why so much? And one would be enough."

Everyone knows x-rays. But not everyone knows that the author of the discovery of x-rays is German physicist Wilhelm Conrad Röntgen. Some of his contemporaries believed that it is indecent to conduct x-ray studies of a person; it is unacceptable (because his skeleton is visible) and therefore demanded the execution of a scientist… Others, on the contrary, proposed to develop theatrical binoculars capable of "seeing through clothes."

Maxwell's great theory began with an article he wrote while he was a student at Cambridge. And of course, no one looked into it. Recognition came with a huge delay. The famous Lord Kelvin (temperature degree was named after him) considered as a hoax the brilliant discovery of x-rays, which for more than a century already serve humanity. He authoritatively declared that devices heavier than air can't fly that radio has no future … French writer Anatole France was right: science is infallible, only scientists are mistaken…

Everyone knows the names of Francis Crick and James Watson, who guessed the magical structure of DNA. Even the discovery of the genetic code often attributed to them. But few people know about George Gamow, who calculated the genetic code, and about Marshall Nirenberg, who opened the way to decoding the code. He took the first step on this path—showed (and of course, accidentally) that synthesis protein polyphenylalanine is encoded by gene, consisting of three uracil-nucleotides. In this case, the Nobel recognition did not have to wait long. But the Russian scientist Boris Belousov (1893–1970), who discovered the famous, stunningly beautiful vibrational modes of chemical reactions, did not live by the date of recognition. He was not believed, his articles were rejected by all the magazines, and reviews were offensive. And only much later it became clear that all this was true, that from Belousov's observations should be coherent chemistry, which obeys all significant biological processes—thinking, heart, muscle, and more. Boris Belousov was graduated from the Swiss Federal Institute of Technology Zurich in 1915 (the same one in which Einstein studied). He was a brilliant, brilliant chemist, a highly educated man, completely devoid of ambition.

After Rutherford's discoveries, journalists once asked Einstein whether it was possible to release the energy predicted by his eq. $E = mc^2$ by bombarding atoms, that is, to "revive" this equation. Einstein's answer was: the discovery of fission of uranium threatens civilization is nothing more than the invention of matches; it is not feasible and is reminiscent of the shooting at the birds in the dark, and in a place where quite a bit birds. And very soon such a terrible "revival" happened in Hiroshima and Nagasaki… Great theater director Nikolai Akimov was right: "making predictions is like that at the birth of a child to write his memoirs, and then force him to live by them."

There are discoveries that change the fate and thinking of mankind. Sir Arthur Stanley Eddington (the one who led the expedition to the Principe Island, who opened the gravitational curvature of light and proved Einstein's theory) noticed that after the discovery of the unimaginable—the wave nature of the electron—"religion has become acceptable to the mind of a reasonable physicist." Erwin Schrödinger wrote about experiments to detect the wave properties of an electron that previously the authors could be placed in a psychiatric hospital to monitor their mental state. Yes, the ways of science are inscrutable and unpredictable…

Einstein, Science and Morality

The person of Einstein is a focus, collecting the characteristics and properties of almost the ideal man of science. As always—seriously and accurately, Einstein defined the status of a scientist: "Belonging to the number of people who give all their strength to reflection and research… is a special honor. I am glad that I have been awarded this honor to some extent, which allows a person to become largely independent of his personal fate and from the actions of others." The great physicist Richard Feynman said that he did not care with whom to talk about physics—with a student or with some lump like the Nobel laureate. This is the inner freedom; this is the aristocracy in science.

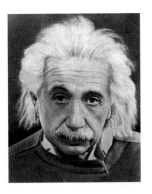

Aristocracy in science is high, unsurpassed professionalism in science, a deep vision and assessment of its state, problems, and prospects, the ability to separate the main from the secondary, without losing sight of either one or the other. This is the ability and skill not only to see the distant horizons and major goals but also to anticipate ravines and failures on the way to those, providing in advance how to overcome or bypass them. It often happens that something conceived as a large scale is in reality a small fake. A sign of aristocracy is not to trifle, not to consider fakes as something significant, not to be attributed to the works and ideas of their students and employees, to get rid of any vanity, and not to use something large to the detriment of others. King Darius (his son Xerxes is known for his command to punish the sea with lashes) defined the universal rules of life: "the most important thing for me is justice. I don't want the weak to suffer the injustice of the strong. I'm no friend of a liar." And he was so in life (died in 486 BC). On the tombstone, he bequeathed to knock out the words: "do not consider perfect what makes the strong; respect what makes the weak."

Einstein was a model of modesty. "I absolutely cannot understand why I was made an idol"—this is from his letter to Max Born. And that wasn't the pose. He was not naive: "the possibilities of knowledge are limitless and the things we have to learn are limitless. But meanness and cruelty also have no boundaries." The greatness of the scientist, according to Einstein, is not his infallibility and perfection; it is his integrity, harmony of mind and conscience, his work of mind on decency. The civil position of justice is also a sign of aristocracy. This does not mean that uncompromising is the property of the aristocrat. Compromises are inevitable, but there is a limit, the critical price of them, which the aristocrat does not step over. Still, Russian poet Maximilian Voloshin warned: "Justice is the multiplication table of the corpses." Because everyone has their own feelings of justice… And of course, the aristocracy is incompatible with haughtiness, arrogance, and false grandeur. Arrogance is a property of the lowest of men. The great Julius Khariton, Scientific Director of the Soviet-Russian nuclear project, was a very modest man. His rule was: striving for the best, not to do the worst. The ability to appreciate and respect their colleagues and their science does not exclude squeamish attitude to the crooks and careerists. The famous phrase of Anatoly Alexandrov (President of the Academy of Sciences of the USSR) is that he knows how to use nuclear energy for peaceful purposes, but does not know how to do it for personal purposes. And then he noticed that there are people who possess such an ability.

Truth, justice, and freedom are Einstein's values. "I always respect the individual and feel an irresistible aversion to violence, ...reject any nationalism, even if it acts as patriotism." But these values were alien to both fascist and communist regimes. Fascism persecuted Einstein, threatened his life (his name was the first in the list of those whom Hitler's gangs intended to destroy) and forced Einstein to leave the country. The communist regime pursued Einstein's name; it was forbidden to mention him in general and especially in scientific articles and books, to make references to his works... They were crossed out by censorship. But scientists were resourceful people... And then in the articles some links to some scientist by the name of Odnokameshkov appeared. Censorship calmly passed it the name of ignorance, but the scientists knew who was hiding behind this name: Odnokameshkov was a translation of the names of Einstein from German to a Russian version of name "*Onestone.*"

During the Nazi persecution, Einstein bitterly and sarcastically remarked: "In dictatorial regimes, the driving force is coercion and lies; in a democracy—only a lie." It seems that he then formulated that people's desire for justice makes democracy possible, but their desire for power makes democracy necessary. It is about such people as Einstein, Goethe put it this way: "Before the great mind I bow my head, before the big heart I kneel."

Einstein formulated the great truth: "the true progress of mankind is based not so much on the ingenuity of the mind, but on the conscience of people." It is a pity that it is forgotten. Einstein echoed Lobachevsky: "intellectual teaching does not a complete education." And finally, another charm of science noticed by Einstein: "scientific research and in general the search for truth and beauty is an area of activity which allows being a child all the life."

Pursuing sciences—complete and all-consuming—affect the scientist's character. They form a calm, wise attitude to life, and disregard for its bustle. It was Einstein who could create such a formula: "anger lives only in the chest of fools." Sounds no worse than $E = mc^2$.

Aesthetics of Science

Science is beautiful and aesthetic in its strength, credibility, and hierarchical logic of subordination of the fundamental components of physics, chemistry, and biology. It has the inner beauty of the mind, which does not need the outer beauty. There's a fascinating mystery about it and it seems to come from the rational beauty of our world. This is only for philosophers "reconciliation of two principles—the scientific and the aesthetic is the most important question facing mankind" (John Dewey, an American philosopher, and psychologist). The question is false, far-fetched, like almost everything in philosophy.

All global theories are excellent and perfect. There is a beauty of finality and inevitability, there is nothing can be changed without destroying it to the ground. Their designs are rigid and stable. They are perfectly accurate in describing the world. The strength of science gives rise even to the suspicions of ignorant people. Paul Feyerabend, the controversial genius of psychology, argued that scientists are monsters who will do anything to prove their theories. But it is the opinion of the philosopher, it is out of science.

Einstein was a man of subtle aesthetics; he realized (not without admiring surprise) that in his theory of relativity the main attractive feature is its logical completeness: none of its conclusions can be refuted. In Newtonian theory, the force of gravity is inversely proportional to the square of the distance between the bodies and this is confirmed by experience. But in it, the quadratic degree can be replaced by a degree of 2.1 or 2.01. Even 1.9 may be used. The theory will not collapse, although it will be less accurate; in this sense, it is semi-empirical. In Einstein's theory, the square of the distance is strictly necessary; it follows from the theory itself, and it cannot be changed—otherwise, the theory will collapse. And it is accurate… When Eddington at the Royal Society reported on the success of his expedition to Principe Island and the discovery of the curvature of light rays—and this was a triumph of Einstein's idea and his theory of relativity—one of the journalists asked Einstein how he would feel if the expedition was unsuccessful and did not find the curvature. He replied: "I would be sorry for God because the theory is correct…" And not the last argument in this confidence was the beauty and aesthetic strength of the theory.

Beauty is the first test of an idea. "There is no place in the world for ugly mathematics" (Godfrey Harold Hardy, an English mathematician). Beautiful quantum mechanics is the most accurate and mysterious science. Someone even said: "Don't touch it, it's beautiful…" It cannot be changed, it is canonical. Moreover, all these fundamental theories are aesthetic and perfect in their mathematical images. Sh. Glashow in the 80s of the last century sharply criticized the theory of strings. In 1997, he said: "we, who worked outside the string theory, have not moved a step... There are questions that will not be answered within the framework of the traditional quantum field theory. This is clear. They may be answered by some other theory, but I don't know any other than string theory." Here again, the aesthetic strength of the theory is not the last argument. Today none of the experienced physicists believe that the standard model is really fundamental one, it is too complex for this; it has many parameters and it is simply ugly.

Undoubtedly, evolutionary theory, linking the evolution of the genotype to evolution of phenotype, is aesthetic one. And even a little aesthetic idea of the struggle for existence as a means of natural selection does not spoil it. First, the idea of struggle is not central to theory; second, new mechanisms of natural selection are emerging, based not so much on struggle as on mutual aid and cooperation within and between species.

We must, however, honestly admit that beyond the framework of fundamental theories the concept of aesthetics becomes unsteady. The classic example of atomic energy is trivial... Information technologies are aesthetic, but "the computer revolution makes it possible to replace educated slaves with ignorant ones" (Russian mathematician Vladimir Arnold). And this is not too aesthetically pleasing... And we are witnesses to that... Science is open, there are thousands of journals to publish what people of science discover. World science functions as a global market of ideas, an open market access to all... What about stealing? The exact answer was given by Russian chemists Victor Kabanov: "in science, it is possible to steal only from the beggar..." One of the heroes of Russian writer Mark Aldanov spoke out: "I forgive people all their shortcomings for their generosity, and I do not put even their virtues for their avarice." The generosity of people of real, high science is a great quality of the scientific community, its high aesthetic indicator.

Science classes and the production of scientific knowledge give aesthetic pleasure. Of course, next to it the torment and desperation of the search area, however, Gaspard Monge was right: "The only charm that accompanies science is able to defeat man's natural aversion to the power of the mind." It has nothing common with conditional reflexes studied by Russian physiologist Ivan Pavlov because not only is the new knowledge promises pleasant but also can threaten with dangers. The pursuit of new knowledge is not a race for pleasure only.

When the speaker tells how he showed that the mechanism of the chemical reaction is complex, I clearly realize that he does not know him. Knowledge is always simple and beautiful. Ahmed Zewail, the creator of femtochemistry, put it this way: "I am sure that behind every meaningful and fundamental concept there must be simplicity and clarity of thought." Aesthetics of science is a reflection of aesthetics and subtle beauty of the world, aesthetics, and beauty of thinking as the main method of cognition of this world. Galileo noticed it four centuries ago, pointing out that science is inscribed on the pages of a huge book, whose name is the Universe, and it is written in the language of mathematics, the most elegant and aesthetically perfect science. Any contradiction, any inconsistency, lack of unity and harmony are unaesthetic.

The contradiction between classical and quantum mechanics was painful for Einstein's aesthetic mind and stubbornly led him to search for a unified theory. The reason is purely aesthetic: in classical physics, there are reasons for everything, there is a certainty; in quantum mechanics there are none. Instead of reliability, there is a probability… The wave function only describes (and does it precisely), but does not explain.

Science is a thing so high and beautiful that it is appropriate to speak about it only in aristocratically refined expressions and subtle linguistic expressions and to write in the language in which the beautiful books of Sir Roger Penrose and Brian Greene were written. The same applies to the people of science; the greatness of the scientist, according to Einstein, is not his infallibility and perfection; it is his integrity, harmony of mind and conscience, his work of mind on decency. By the way, this applies to any person and to any profession… "The true progress of mankind is based not so much on the ingenuity of the mind, but on the conscience of people"—this is the phrase of Albert Einstein. And finally, another charm of science noticed by Einstein: "Scientific research and in general the search for truth and beauty is an area of activity in which it is allowed to remain a child all his life."

Absolute Truth: Virtual Science

Yes, there are absolute truths… And there are few of them.

1. The laws of conservation of energy and angular momentum are absolute. If someone promises to make an engine with 100.1% efficiency or even 200 and 1000%, or create a machine that converts vacuum energy into electricity, ignore those promises.
2. No speed can exceed the speed of light. (Recently, there have been reports of exceeding this speed. Isn't that a new fusion in the tin?)
3. The Universe is expanding.

4. "The true progress of mankind rests not so much on the ingenuity of the mind as on the conscience of the people" (Einstein).
5. The temperature cannot be below absolute zero (-273.16 °C).

The speed of light is the absolute truth, but the mass of an electron is not: it depends on the speed of its movement. Virtual science is not a curse or a reproach. This is real science, but its goals are either unattainable (the pursuit of artificial intelligence or earthquake prediction), or nondescript, insignificant, and even false... Earthquake is the phenomenon of pulse discharge of elastic energy stored in the area of the earth's crust, called the hearth. The earthquake source is a giant physical-mechanical and mechano-chemical macro-reactor, where the events are determined by two competing processes: the accumulation of energy due to deforming forces (deformation energy pumping) and its discharge through a catastrophic shift, the sliding of some parts of the Earth's crust relative to others; this shift brings innumerable disasters. Both processes are inevitable and natural for a living, dynamic system like the Earth. Another intriguing factor is the huge division, the incommensurability of time and space scales of these two processes. Speeds of movement of sites of the Earth crust providing energy pumping of the center make centimeters or even millimeters a year, and energy dumping occurs at the movement of the Earth sites with the speeds reaching meters per second; a difference in speeds makes more than ten orders. This fact is reflected in the frequency of earthquakes generated by the same source: usually, the time of energy accumulation, i.e., the period of preparation, "silence" of the source lasts for decades (and even centuries), while the release of energy occurs in seconds (maximum—in minutes).

The problem of earthquakes is global, universal, and eternal. It is believed that the impact on the earthquake as a monumental phenomenon is impossible and therefore need to look for signs of its approach—the harbingers of the earthquake, to predict the time of the disaster to take security measures. The science of the forecast was formed around this point of view.

Numerous forecasts of the last 100 years have been unsuccessful and erroneous. American seismologist and physicist Charles Francis Richter, whose name is called the scale of earthquakes, said that he "does not like the pathological interest in the forecast." The point here, however, is not in taste—the price of the problem is too high. International Association for seismology and physics of the Earth with the participation of physicists of all countries have carried out fundamental work on the analysis of all forecasts and all of the signs, referred to as precursors. Its outcome: none of the signs is satisfactory in reliability; they often contradict each other; none of them is effective.

And there are strict physical grounds for that judgment. Earthquake source is an open dynamic system, where the boundary between stationary and non-stationary (explosive) regimes is determined by critical conditions that are unpredictable, uncontrolled and therefore uncontrollable. Due to the strong nonlinearity of the system, it is uncontrollable: infinitesimal instabilities can stimulate large ones. It is impossible to take into account all random and infinitely small disturbances as well as "it is impossible to live in detail" (Lev Tolstoy). In this sense, the whole science of forecasting is a virtual one, its goal is unattainable.

The earthquake source is not the only system of this kind. It is impossible to predict the moment of explosion of a flask with an explosive mixture of $2H_2 + O_2$, although all reactions and their velocities in this system are known with high accuracy. We do not control the transition from laminar to turbulent flow, from combustion to detonation, as well as social explosion (revolution), etc. It is impossible to predict the unpredictable... The weather forecast and all meteorology belong to the same science. Academician A. N. Krylov equated meteorology to palmistry. It's possible to equate it to astrology too...

The author considers the synthesis of superheavy elements (those that follow dubnium, nilsbohrium, etc.) as an example of science with minor goals. It is not obvious sense to get a few unstable, short-lived atoms of a new element, the properties of which are known in advance from the Mendeleev periodic table of elements. It is possible, of course, to reach the island of nuclear stability (its existence is reliably predicted by the theory of nuclear forces), but the price of these unnecessary elements will be fantastic. Very smart and decent people work in this science, but it would be wiser to redirect the funds sent to this virtual project, for example, to send to biology and medicine.

The project of the flight to Mars belongs to the same type. The peculiarity of Russian science is that its management system copies the system of government. It is referred to as a managed democracy (formerly called dictatorship). In such a system, scientific priorities and projects are appointed by supervisors; and then the appropriateness criteria are irrelevant. But people are not interested in whether there is life on Mars, they are much more interested in health and well-being here on Earth. And the health of the Earth...

Deception, Pseudoscience, and Sensationalism

Science is being alive and quivering. There are losses and finds, ups and downs, avenues and deadlocks, ideas and prejudices, truth and deception on the roads of science. "The difficulty lies not in generating new ideas, but in saving from old ones" (John Maynard Keynes, an English economist). After all, prejudice—it is a piece of the old truth (Yevgeny Baratynsky, a Russian poet), ceased to be true.

Science is a system with self-cleaning, it gets rid of the old truth, replacing it with another one—updated and refined. Here are examples of prejudices, the wreckage of the old truth. Albert Einstein: "If you can't measure anything, you don't understand it." Yes, quantum mechanics is devoid of common sense, it is not given to understanding. But it is absolutely accurate, perfectly measurable, and strictly quantitative. Jacob Berzelius: "Who issues likely to be true, he is a deceiver." But in quantum mechanics, the probable is the true…

Some Joker started the idea that the wheel is the greatest discovery of mankind. The idea is sustainable, a survivor of the century… Although there was no essential discovery: even the wild ancestors of humans knew that all-round easier to roll than carrying or drag. However, the idea is eternal and mindlessly repeated… Thoughtlessness, like madness, is incurable. And there are many such "wheels" in science and in life. But there is another example of "the wheel" that is exquisite and highly intelligent—the anthropic principle. Its meaning is in the statement that the laws of nature and the structure of the world are remarkably adapted "for life," for the existence of the living world and man. One has only a little "move," change the world constants—Planck constant, the constant of gravity, the speed of light, the mass of the electron, etc.—then the world and its most vulnerable part—life—will collapse.

It is known that the living world is built on a carbon foundation and this, according to the anthropic principle, is not accidental. Within 3 min after the Big Bang, when the temperature dropped from 10^{32} degrees to a billion degrees, protons and neutrons began to unite into atoms (more precisely, in the nuclei of chemical elements). First helium was born: two protons and two neutrons merged. Then two helium nuclei are merged into a nucleus of beryllium-8; then the nucleus ^{8}Be has attached a helium nucleus and it turned into a nucleus of carbon-12. But in order for carbon to be abundant so that it accumulates in nuclear mergers, there must be such an energy level in the ^{12}C nucleus that its energy is close to the sum of the energies of the ^{8}Be and ^{4}He nuclei. This level exists and was found experimentally. It is a successful combination of physical constants of nuclear interactions that ensured the stability of the level and the appearance of carbon in an amount sufficient for the appearance of life. There would be no carbon—it is possible that there would be another life...

The anthropic principle is the illusion that all physical and cosmological processes were arranged and organized to create life at the finish. This is a false principle: the Big Bang and the whole cosmological history developed according to their physical laws, which were not prudent and did not care about a future life. On the contrary, life itself appeared, built into the ready world, adapted to it, to its parameters and its conditions. The anthropic principle is a brilliant joke of physicists, thrown to philosophers. For them, this gift is a boundless space for wisdom; they believe that it is a giant philosophical and religious values that hides the truth. Their thesis: the Universe must be arranged in such a way as to allow the existence of beings with our bodily organization. And someone had to take care of it...

And there is another "if" of the anthropic principle. If the difference of the masses of the neutron and proton were a little bit smaller than the one that is (1 MeV), the more sustainable would be a neutron, not a proton. And then the hydrogen atom would be unstable, the synthesis of elements and our organic world could not be realized. Such "if" is filled with the world, science knows them. And only through science comes a deep and meaningful understanding of the uniqueness of the world, which was born accidentally and in which we were also born accidentally...

Around science is always an aura of mystery. Thus many adventurers and ignorant wind around (and fed), speculating that halo. Sometimes pseudoscience is sincere (due to ignorance), but sometimes it is artificial due to cunning. Both are popular: the number of palmists, psychics, ufologists, atlantologists (seekers of mythical Atlantis), fortune-tellers, and mystics have already surpassed the number of scientists. (The Academy of Sciences issues a Yearbook "In defense of science"; it colorfully describes all the miracles that imitators promise, the most dangerous for both science and society people. This can be tolerated and condescended as a mind game, but if it is portrayed by science, patience is misplaced. Myths are not science.) The fight against pseudoscience is respected, but almost hopeless and doomed to failure. It is like a war with windmills, which was led by a well-known naive and beloved hero Don Quixote. Because there are financial interests behind pseudoscience that derived from good science. For example, stupidity under the sonorous name of quantum medicine fabulously enriches its enthusiasts.

The reason for the prosperity of pseudoscience is obvious: the development of true and sound knowledge requires at least a little stress of the mind, while fakes on behalf of science are brought ready, carried away by false promises and trustfully perceived by society. "Ignorance is the night of the mind" (Cicero). It so happened that science has become a hostage of its power and authority. It's that exotic case where the vanguard of science is at the back... And it's hardly a joke.

Yes, there is cunning in science... But it is lying rarely (see below), it is just not telling the truth. And sometimes it triggers myths (similar to tales of global warming or the turn of the Gulf Stream and global glaciation). Often these myths are a way to extract money from stupid governments, for example, to promise the impossible, to predict an earthquake or to create a car with a water-powered engine.

In science, there may be a direct deception, manipulation of results, and false discoveries. Most often it happens in archaeology, biology, and medicine. Such histories in natural sciences are exceptions; they are rare and usually unfold sooner or later. There is an article entitled "Deception in science" in the Russian peer-reviewed scientific journal *Physics-Uspekhi* (Advances in Physical Sciences), 1993, №1. Critics of science (those are mostly philosophers) seek (and find!) elements of deception even in the scientific writings of Newton, Galileo, and other giants of science. This deception is almost "sinless," it is associated with discarding the results of some experiments, with the selection of them in order to add credibility to their work. So, a case relating to the work of the great Gregor Mendel is known. Austrian biologist and botanist, Augustinian monk Gregor Johann Mendel discovered the laws of genetics. Mathematician R. Fisher proved that the quantitative data, which led the "great monk" to prove his laws, could not be obtained. But the laws of genetics remained and they are accurate. Another historian of science examined the working diaries of the great American experimental physicist Robert Andrews Millikan who measured the charge of the electron on the motion of charged oil drops in an electric field. He found that the Millikan discard some results as false and explained the reasons for this selection. Many experimenters do this and it makes sense. Ultimately, people are not interested in the scientist's cuisine; only the final results are important, not the search, but the findings.

Along with real science, there is pseudoscience, i.e., imitation of scientific research and mimicry for science. Pseudoscience presents senseless as something significant (for example, the forecast of the economy for 50 years ahead, that the current scientists-economists are fond of). A sign of pseudoscience is false ambiguity. There is a shadow science (such as the shadow economy): senseless scientific papers, not containing any scientific information. There is Ghost, positioning himself in science, only the name of the Director and sign; an essay of fictional results; the production of scientific information garbage.

The harmless pseudoscience has done numerous "discoveries." Those "discoveries" are usually set out in the letters of dabblers. Recently there was a "new fundamental science" that was called organismic, declared as a global central science. It has its own laws and ridiculous theorems. Here is one of them: the sum of two numbers, equal in modulus, but having opposite signs, is not zero. No comment, as they say over the ocean…

The most intriguing thing in science is a sensation. Most of them come from chemistry. They blow up the whole intellectual world, which is much wider than the scientific world. One of the recent sensations was the discovery of substances with high-temperature superconductivity. Nothing significant came of it, but the sensation was not false: chemistry, physics, and technology advanced to new horizons of knowledge. The latest sensation was the creation of graphene. It is very likely that here, too, expectations are greatly exaggerated and the final results will be more than modest.

But in the modern history of science, there was a great sensation, bursting like a soap bubble. In this sensation, everything was unusual. The content itself could blow anyone up: it was a high-speed nuclear reaction at room temperature. Here is this reaction (it goes on two channels):

$$D + D \rightarrow T + p$$

(here the fusion of two deuterium nuclei gives rise to a heavy hydrogen isotope tritium T and proton p),

$$D + D \rightarrow He + n;$$

in this channel the fusion of deuterium nuclei produces helium He and a neutron n.

These reactions are a source of tremendous energy. Science has been struggling for decades to make these reactions come true and create a source of thermonuclear energy through the thermally stimulated deuterium nuclear reaction. This requires temperatures of hundreds of millions of degrees. And when it was reported that these reactions were easy to go in the bank of jam and any housewife in the kitchen can carry out—it is a shock…

And it was a shock when two chemists, Martin Fleischman and Stan Pons, announced that by passing an electric current through palladium with hydrogen dissolved in it, a huge amount of energy was released, neutrons and helium, tritium, protons, and even gamma radiation appeared. Unexpected was the fact that such processes (this is common electrochemistry) have been well studied and no one has never seen. The form of presenting the results was unusual: through interviews with newspapers and television, rather than by the standard way in science—the publication of an article in a peer-reviewed scientific journal.

The world was divided: some went into complete disbelief, others enthusiastically rushed to repeat the experiments. Provincial and little-known University of Utah where it was first reported about the "cold fusion" (so they began to call this phenomenon) immediately became world famous. Instantly there was a stream of evidence of effectiveness. It was reproduced in some agricultural Institute of Japan, in Kharkiv Institute of physics and technology, in Kiev Institute of nuclear physics, at the physics faculty of Moscow State University. It was confirmed in Hungary. The German Democratic Republic (was once such) reported success through its Secretary-General Erich Honecker. Vienna and Tokyo reported their findings. On 6 April 1989, the government of Utah announced the grant five million dollars for the work of Fleischmann in the University of Utah. The Dean of The Southampton School of physics told ten labs to replicate the results. The Italians announced a world conference on cold fusion in mid-April. The thermonuclear theme became the main one on the first pages of newspapers and TV programs. Here is an excerpt from the report of the Institute of chemical physics of the USSR as of April 24, 1989, signed by the Director academician Vitaliy Goldansky and presented to the Academy of Sciences: "the Institute of chemical physics of the USSR as delivered the first experiments to verify reports of "low-temperature" nuclear synthesis. In the near future, new experiments are planned... In the electrolysis of deuterated water with a palladium electrode, 1.5–2 h after the current was turned on, a three-fold excess of the neutron flux over the background was recorded. There was no effect in the electrolysis of usual water. Currently improving the system of the account of neutrons is preparing to work the counter of tritium."

The Russian Academy of Sciences had a special Commission (the author of these notes was its member), which collected all the information and evaluated science. Its Chairman and Vice-President academician Oleg Nefedov took a smart and restrained position. At one of the meetings in the Commission, the representative of the Atomic Energy Committee of the USSR said that all the effects are artifacts that actually do not exist. But this did not make an impression; the statement was taken as a sign of professional jealousy. The voices of professional scientists who pointed to possible sources of errors deserved more attention. But they were drowned in the euphoria of delight from the extravagance and scale of ideas and phenomena.

In the same April of the same year, the annual and traditional Congress of The American Chemical Society was held in Dallas. The author of these notes was present in a huge hall similar to an indoor stadium. There were more than 10,000 people and there was not a normal indifferent. Curiosity, distrust, and sarcasm were found in the smiles and remarks that my neighbors exchanged. There was barely visible tiny chair at the end of the room, where Fleischman and presiding were sitting. They said something, but they hardly listened; they were all busy talking among themselves.

Almost simultaneously with Fleischman and Pons, American geophysicist S. Jones presented his results on neutrons, γ-radiation, and heat in palladium and titanium electrochemistry at the Congress of the American Physical Society on May 1989. He admitted that he had observed neutrons and γ-quanta, but the heat was negligible. Fleischmann and Pons reported that electrolysis generated 14 W/cm^3 of heat, and Jones got the heat up to 10^9 times smaller. "It's like a one dollar piece of paper in relation to the US Federal budget," he said. Physicists noticed another thing: if the heat measured by Fleischman and Pons would correspond to the number of neutrons, then the experimenters would have been dead from the terrible power neutron flux.

Physicists of Harwell Laboratory (England) preserved sobriety of mind. On April 7, 1989, at the conference, they stated that the results of Fleischmann and Pons were not reproduced. Recall also that in metals (including palladium, from which the electrodes were made), there is always helium-3 and helium-4. Their detection was no evidence of cold fusion. The finish came on August 1989 in Stockholm at 32 Conferences of the International Union of Pure and Applied Chemistry (IUPAC). There was also Fleishman with his wife, but there was a complete disappointment and bitter indifference to the unfulfilled, that seemed so charming…

Why did this happen? It is not necessary to think that all participants in this performance were ignorant or swindlers. Everyone knew that for the thermonuclear reaction $D + D \rightarrow {}^3He + n$ (fusion of two deuterium nuclei with the generation of helium-3 and neutron) it is required to overcome the energy barrier of 60,000 electron volts. And 1000 electron volts corresponds to the thermal energy of 11.5 million degrees. So, for the fusion of deuterium nuclei need to accelerate them to speeds that are achieved at 700 million degrees. For electrochemical experiments, it is excluded. But in the crystal lattices of palladium (electrodes are made of it), there are internal pressures that compress the deuterium ions embedded in the lattices. It can be expected that this pressure will "squeeze" deuterium nuclei into each other, overcoming the giant barrier of their Coulomb repulsion. However, the internal pressure is amenable to calculation and it was known that it is not enough to overcome the barrier.

In addition, in the scientist subconscious a joke of Einstein alive: "Everyone knew that this cannot be done. One person didn't. He made a discovery." Unlike smart and knowledgeable, the ignorant do not know that something should not be done. They are doing and accidentally open. It is pity, it didn't happen in the cold fusion. How did it end? Nothing... In this case, only singles enthusiasts remained; serious scientists have an opinion about them, which intelligent people do not express aloud...

Science, Power, and Hypocrisy of Progress

Science is an elite part of civilization, its highest culture. It is majestic and eternal, like the pyramids of Egypt. Extraction of knowledge is a harmless thing, so true science is pure and flawless. Opening new knowledge, it performs two functions—creates useful things and detects things dangerous to people, warning about the dangers. Knowledge itself is a power; this phrase is translated into Russian as "knowledge is a force" and in this sound it became winged. But the first true translation is "knowledge—power," and the second—force.

But the government does not like competition. It likes to be the one. Power is dominated by the people of vanity, and vanity is proportional to their limitations. Such people are afraid of smart people; do not like free minds and independent characters. This is at best; at worst, repressed science emerges. In the twentieth century, the Nazis burned books, persecuted Einstein and destroyed the best, the "Jewish" science. Communist Bolsheviks also banned Einstein, smashed cybernetics, genetics, biology, and chemistry. But they went on: they had an executed science.

All modern strong states have strong and highly developed science. Without science, their helplessness and state disability are revealed. Great minds are aware of this and instructed (often fruitless) their rulers. Vladimir Vernadsky: "a nation that does not know how to properly use its greatest wealth—the mental strength of its workers—inevitably suffers and lags behind in the world life competition." Napoleon Bonaparte in August 1812, already from Russia, from Vitebsk, wrote Laplace: "the propagation and improvement of science are closely connected with the welfare of the state." Even this owner of Europe understood something... Ahmed Zevail, Nobel laureate, member of the Russian Academy of Sciences: "the whole history convinces that Nations degrade if their leaders do not respect science and education." Doing science and taking care of science means investing in the future. Those who do not such investing will have no future. This is a simple and irrefutable truth.

Science is at the edge of progress, i.e., creation of the artificial world adapted to the egoism of man and mankind. In this world, there are many beautiful, but even more dangerous and ugly, as far as it is created at natural world expense and its destruction. This is the iron pace of progress and this is its hypocrisy. A wonderful and sincere writer Yuri Nagibin naively noted in his "Diaries": "It looks wildly, to be born in the country house with candlelight and kerosene lamps in the villages—to die from a neutron explosion. Science is developing too fast." The hypocrisy of progress is living on credit; we robbed future generations of many, hurt and devastated them. This loan will never be repaid. The hypocrisy of progress is the excessive comfort of the wealthy, paid for by poverty and the suffering of billions of people. Such advantage was not created by labor and talent, but by robbery and embezzlement. The hypocrisy of progress is a false scale of assessments: to know the price of everything but to ignore the values. And this is a devastating disease of civilization, behind which is the loss of self-preservation and doom. Any calls to realize this danger is naive and empty. Diseases of civilization are hidden deeply in the genes. The egoism is universal, altruism has been sporadic, and it always shyly concedes to egoism. Genetic egoism destroys the genetic unity of mankind and turns into the hypocrisy of humanism. Humanity is doomed to rapid progress, but its hypocrisy is fraught with death. Democracy—the highest achievement of progress—is hypocritical: we choose servants who always turn out to be our masters.

What about science? She's on the cutting edge. The place is ambiguous… Physics has mastered the power of the atom and taught mankind to take it. There were nuclear power plants, and nearby there is the weapon of mass destruction of people. The miracle of science has become a monster of life. The light of knowledge became a torch for arson. Chemistry gives mankind a lot of benefits and means of treating illness, but near there are toxic substances. Both the atomic bomb and the toxic substances have been tested and their victims are enormous… Biology brings to mankind deliverance from diseases and many other benefits. However, bacteriological weapons and unprecedented dangers of genetic engineering are the next…

What to do? Philosophers think that it is necessary to stop science. But, first, it is naive because it is impossible. Secondly, the dangers do not come from science, it only creates benefits, but other forces decide their fate and distribute. The dangers and hypocrisy come from the use of science, from the address and goals to which its discoveries and conquests are directed. And we must not put blind obstacles in the way of progress, but put obstacles in the way of blind progress. Science is not the creator of good or evil, its job is to study the world, its properties, to turn the secret into the obvious, unpredictable into the inevitable. It is the duty of society to take care of the fate of scientific discoveries and the correct addresses of their use. Where is it? Recall the will of the wise Einstein: "the true progress of mankind is based not so much on the ingenuity of the mind, but on the conscience of people"…

Science Is a Festive of Business

Of course, science is available to gloomy minds, but it is more favorable to bright and sharp minds. Russian mathematician Nikolai Lobachevsky noticed: "For those who have the mind dulled and the sense deaf, nature is mort, alien beauty of poetry is alien, the charm of the architecture is devoid of, and a story of the century is not interesting." We can add: they do not hear the music of quantum strings, chemical and genetic notes, the poetry of molecules, the magic of mathematics and quantum mechanics.

Scientific creativity is a joyful and festive business. It is serious but decorated with a joke. Predominantly physicists (remember the popular book "Physicists joke," published three times—in 1965, 1968, and 1993), mathematicians and chemists joke (see the wonderful book by Y.A. Zolotov, "Chemists are still joking," 2008; M. G. Voronkov and A. Yu Rulev "About chemistry and chemists in jest and earnest," 2011). Biologists joke a little. Because they deal with life. But jokes with life are dangerous…

Good humor in science is peculiar; it is distinguished by aristocracy, sophistication, and the game of professional mind. One of the heroes of the Polish writer Ch. Khrushchevsky noted: "humor reveals the heaviest gate that closes access to wisdom." The great Pyotr Kapitsa argued that "science should be fun, exciting and simple. Those must be the scientists." Here are a few examples of humor from the book of M. G. Voronkov and A. Rulev.

At the chemistry lesson, the teacher asked what would happen to gold if it would lie in the open air (the property of gold was implied not to be oxidized). The student answer was: "It will disappear."

How do great scientists differ from others? The first give life to science, the second receive from science for life.

© Springer Nature Singapore Pte Ltd. 2020
A. L. Buchachenko, *The Beauty and Fascination of Science*,
https://doi.org/10.1007/978-981-15-2592-6_11

Here's a joke by a brilliant German chemist Heinrich Wieland. At the time of Nazism, in one of his lectures, he said that the brain contains phosphorus, and added: "Today Germany is the poorest country with phosphorus."

There is remarkable professional self-irony. Justus Liebig, creating a simple method of analysis of chemicals, delightfully said: "now a monkey can be a chemist." Adolf Bayer, a brilliant chemist, willingly told everyone that he became a chemist because someone told him that a chemist needed to think only once a year.

And the word "chemistry" goes back to the words "kidnap," "steal," "steal"… and the authors conclude that a chemist is a clever man, skillfully stealing from nature its secrets and gives them to serve the people. And it's a great service…

IIII IIIIIIIIIIIIIIIIIIIII

Printed in the United States
by Baker & Taylor Publisher Services